Biostatistics and Epidemiology
Third Edition

Springer

New York
Berlin
Heidelberg
Hong Kong
London
Milan
Paris
Tokyo

Sylvia Wassertheil-Smoller

Professor and Head, Division of Epidemiology,
Department of Epidemiology and Population Health
Holder of the Dorothy Maneleoff Foundation
and Molly Rosen Chair in Social Medicine
Albert Einstein College of Medicine

Biostatistics and Epidemiology

A Primer for Health and Biomedical Professionals

Third Edition

With 22 Illustrations

 Springer

Sylvial Wassertheil-Smoller
Department of Epidemiology and Population Health
Division of Epidemiology
Albert Einstein College of Medicine
Bronx, NY 10461-1602
USA
smoller@aecom.yu.edu

Library of Congress Cataloging-in-Publication Data
Wassertheil-Smoller, Sylvia
    Biostatistics and epidemiology : a primer for health and biomedical prefessionals / by Sylvia
Wassertheil-Smoller.—3rd ed.
        p.   cm.
    Includes bibliographical references and index.
    ISBN 0-387-40292-6 (alk. paper)
    1. Epidemiology—Statistical methods. 2. Clinical trials—Statistical methods. I. Title.
RA652.2.M3W37   2003
614.4'072—dc22   2003058444

Printed on acid-free paper.

ISBN 0-387-40292-6   Printed on acid-free paper.

Printed in the United States of America.   **(BPR/MVY)**

9 8 7 6 5 4 3 2 1   SPIN 10935774

Springer-Verlag is a part of *Springer Science+Business Media*

springeronline.com

To Jordan

PREFACE TO THE THIRD EDITION

and summary of prefaces to the first two editions

This book, through its several editions, has continued to adapt to evolving areas of research in epidemiology and statistics, while maintaining the original objective of being non-threatening, understandable and accessible to those with limited or no background in mathematics. Two new areas are covered in the third edition: genetic epidemiology and research ethics.

With the sequencing of the human genome, there has been a flowering of research into the genetic basis of health and disease, and especially the interactions between genes and environmental exposures. The medical literature in genetic epidemiology is vastly expanding and some knowledge of the epidemiological designs and an acquaintance with the statistical methods used in such research is necessary in order to be able to appreciate new findings. Thus this edition includes a new chapter on genetic epidemiology as well as an Appendix describing the basics necessary for an understanding of genetic research. Such material is not usually found in first level epidemiology or statistics books, but it is presented here in a basic, and hopefully easily comprehensible way, for those unfamiliar with the field. The second new chapter is on research ethics, also not usually covered in basic textbooks, but critically important in all human research. New material has also been added to several existing chapters.

The principal objectives of the first edition still apply. The presentation of the material is aimed to give an understanding of the underlying principles, as well as practical guidelines of "how to do it" and "how to interpret it." The topics included are those that are most commonly used or referred to in the literature. There are some features to note that may aid the reader in the use of this book:

(a) The book starts with a discussion of the philosophy and logic of science and the underlying principles of testing what we believe against the reality of our experiences. While such a discussion, per se, will not help the reader to actually "do a t-test," I think it is important to provide some introduction to the underlying framework of the field of epidemiology and statistics, to understand why we do what we do.

(b) Many of the subsections stand alone; that is, the reader can turn to the topic that interests him or her and read the material out of sequential order. Thus, the book may be used by those who need it for special purposes. The reader is free to skip those topics that are not of interest without being too much hampered in further reading. As a result there is some redundancy. In my teaching experience, however, I have found that it is better to err on the side of redundancy than on the side of sparsity.

(c) Cross-references to other relevant sections are included when additional explanation is needed. When development of a topic is beyond the scope of this text, the reader is referred to other books that deal with the material in more depth or on a higher mathematical level. A list of recommended texts is provided near the end of the book.

(d) The appendices provide sample calculations for various statistics described in the text. This makes for smoother reading of the text, while providing the reader with more specific instructions on how actually to do some of the calculations.

The aims of the second edition are also preserved in this third edition. The second edition grew from feedback from students who indicated they appreciated the clarity and the focus on topics specifically related to their work. However, some users missed coverage of several important topics. Accordingly, sections were added to include a full chapter on measures of quality of life and various psychological scales, which are increasingly used in clinical studies; an expansion of the chapter on probability, with the introduction of several nonparametric methods; the clarification of some concepts that were more tersely addressed in the first edition; and the addition of several appendices (providing sample calculations of the Fisher's exact test, Kruskal–Wallis test, and various indices of reliability and responsiveness of scales used in quality of life measures).

It requires a delicate balance to keep the book concise and basic, and yet make it sufficiently inclusive to be useful to a wide audience. I hope this book will be useful to diverse groups of people in the health field, as well as to those in related areas. The material is intended for (1) physicians doing clinical research as well as for those doing basic research; (2) for students—medical, college, and graduate; (3) for research staff in various capacities; and (4) for anyone interested in the logic and methodology of biostatistics and epidemiology. The principles and methods described here are applicable to various substantive areas, including medicine, public health, psychology, and education. Of course, not all topics that are specifically relevant to each of these disciplines can be covered in this short text.

Bronx, New York   Sylvia Wassertheil-Smoller

ACKNOWLEDGMENTS

I want to express my gratitude for the inspired teaching of Dr. Jacob Cohen, now deceased, who started me on this path; to Jean Almond, who made it personally possible for me to continue on it; and to my colleagues and students at the Albert Einstein College of Medicine, who make it fun.

My appreciation goes to those who critiqued the first and second editions: Dr. Brenda Breuer, Dr. Ruth Hyman, Dr. Ruth Macklin, Dr. Dara Lee, Dr. C.J. Chang, and Dr. Jordan Smoller, as well as the many students who gave me feedback. My thanks to Dr. Gloria Ho, Dr. Charles Hall, Dr. Charles Kooperberg, Mimi Goodwin, and Dr. Paul Bray for help in editing new material in the third edition and for their helpful suggestions.

Very special thanks go to my son, Dr. Jordan Smoller, for his invaluable help with the material on genetics, his thorough editing of the book—and for everything else, always.

I am greatly indebted to Ms. Ann Marie McCauley, for her skillful and outstanding work on the earlier editions and to Ms. Darwin Tracy for her artistic preparation of the new manuscript, her remarkable patience, commitment, and hard work on this book.

Finally, my deep love and gratitude go to my husband, Walter Austerer, for his help, encouragement, and patience.

Bronx, New York   Sylvia Wassertheil-Smoller

CONTENTS

Chapter 1
THE SCIENTIFIC METHOD

Science is built up with facts, as a house is with stones. But a collection of facts is no more a science than a heap of stones is a house.

Jules Henri Poincare
La Science et l'Hypothese (1908)

1.1   The Logic of Scientific Reasoning

The whole point of science is to uncover the "truth." How do we go about deciding something is true? We have two tools at our disposal to pursue scientific inquiry:

We have our senses, through which we experience the world and make *observations.*

We have the ability to reason, which enables us to make logical *inferences.*

In science we impose *logic* on those observations.

Clearly, we need both tools. All the logic in the world is not going to create an observation, and all the individual observations in the world won't in themselves create a theory. There are two kinds of relationships between the scientific mind and the world, two kinds of logic we impose—*deductive and inductive,* as illustrated in Figure 1.1.

In *deductive inference,* we hold a theory and based on it we make a prediction of its consequences. That is, we predict what the observations should be. For example, we may hold a theory of learning that says that positive reinforcement results in better learning than does punishment, that is, rewards work better than punishments. From this theory we predict that math students who are praised for their right

1

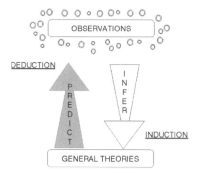

Figure 1.1

answers during the year will do better on the final exam than those who are punished for their wrong answers. We go from the general—the theory—to the specific—the observations. This is known as the hypothetico-deductive method.

In *inductive inference,* we go from the specific to the general. We make many observations, discern a pattern, make a generalization, and infer an explanation. For example, it was observed in the Vienna General Hospital in the 1840s that women giving birth were dying at a high rate of puerperal fever, a generalization that provoked terror in prospective mothers. It was a young doctor named Ignaz Phillip Semmelweis who connected the observation that medical students performing vaginal examinations did so directly after coming from the dissecting room, rarely washing their hands in between, with the observation that a colleague who accidentally cut his finger while dissecting a corpse died of a malady exactly like the one killing the mothers. He inferred the explanation that the cause of death was the introduction of cadaverous material into a wound. The practical consequence of that creative leap of the imagination was the elimination of puerperal fever as a scourge of childbirth by requiring that physicians wash their hands before doing a delivery! The ability to make such creative leaps from generalizations is the product of creative scientific minds.

Epidemiologists have generally been thought to use inductive inference. For example, several decades ago it was noted that women seemed to get heart attacks about 10 years later than men did. A creative leap of the imagination led to the inference that it was women's hormones that protected them until menopause. EUREKA! They deduced that if estrogen was good for women, it must be good for men and predicted that the observations would corroborate that deduction. A clinical trial was undertaken which gave men at high risk of heart attack estrogen in rather large doses, 2.5 mg per day or about four times the dosage currently used in post-menopausal women. Unsurprisingly, the men did not appreciate the side effects, but surprisingly to the investigators, the men in the estrogen group had higher coronary heart disease rates and mortality than those on placebo.[2] What was good for the goose might not be so good for the gander. The trial was discontinued and estrogen as a preventive measure was abandoned for several decades.

During that course of time, many prospective observational studies indicated that estrogen replacement given to post-menopausal women reduced the risk of heart disease by 30-50%. These observations led to the inductive inference that post-menopausal hormone replacement is protective, i.e. observations led to theory. However, that theory must be tested in clinical trials. The first such trial of hormone replacement in women who already had heart disease, the Heart and Estrogen/progestin Replacement Study (HERS) found no difference in heart disease rates between the active treatment group and the placebo group, but did find an early increase in heart disease events in the first year of the study and a later benefit of hormones after about 2 years. Since this was a study in women with established heart disease, it was a secondary prevention trial and does not answer the question of whether women without known heart disease would benefit from long-term hormone replacement. That question has been addressed by the Women's Health Initiative (WHI), which is described in a later section.

The point of the example is to illustrate how observations (that women get heart disease later than men) lead to theory (that hormones are protective), which predicts new observations (that there will be fewer heart attacks and deaths among those on hormones), which may strengthen the theory, until it is tested in a clinical trial which can either corroborate it or overthrow it and lead to a new theory, which then must be further tested to see if it better predicts new observations. So there is a constant interplay between inductive inference (based on observations) and deductive inference (based on theory), until we get closer and closer to the "truth."

However, there is another point to this story. Theories don't just leap out of facts. There must be some substrate out of which the theory leaps. Perhaps that substrate is another preceding theory that was found to be inadequate to explain these new observations and that theory, in turn, had replaced some previous theory. In any case, one aspect of the "substrate" is the "prepared mind" of the investigator. If the investigator is a cardiologist, for instance, he or she is trained to look at medical phenomena from a cardiology perspective and is knowledgeable about preceding theories and their strengths and flaws. If the cardiologist hadn't had such training, he or she might not have seen the connection. Or, with different training, the investigator might leap to a different inference altogether. The epidemiologist must work in an inter-disciplinary team to bring to bear various perspectives on a problem and to enlist minds "prepared" in different ways.

The question is, how well does a theory hold up in the face of new observations? When many studies provide affirmative evidence in favor of a theory, does that increase our belief in it? Affirmative evidence means more examples that are consistent with the theory. But to what degree does supportive evidence strengthen an assertion? Those who believe induction is the appropriate logic of science hold the view that affirmative evidence is what strengthens a theory.

Another approach is that of Karl Popper, perhaps one of the foremost theoreticians of science. Popper claims that induction arising from accumulation of affirmative evidence doesn't strengthen a theory. Induction, after all, is based on our belief that the things unobserved will be like those observed or that the future will be like the past. For example, we see a lot of white swans and we make the assertion that all swans are white. This assertion is supported by many observations. Each time we see another white swan we have more supportive evidence. But we cannot prove that all swans are white no matter how many white swans we see.

On the other hand, this assertion can be knocked down by the sighting of a single black swan. Now we would have to change our assertion to say that most swans are white and that there are some black swans. This assertion presumably is closer to the truth. In other words, we can refute the assertion with one example, but we can't prove it with many. (The assertion that all swans are white is a descriptive generalization rather than a theory. A theory has a richer meaning that incorporates causal explanations and underlying mechanisms. Assertions, like those relating to the color of swans, may be components of a theory.)

According to Popper, the proper methodology is to posit a theory, or a conjecture, as he calls it, and try to demonstrate that it is false. The more such attempts at destruction it survives, the stronger is the evidence for it. The object is to devise ever more aggressive attempts to knock down the assertion and see if it still survives. If it does not survive an attempt at *falsification*, then the theory is discarded and replaced by another. He calls this the method of *conjectures and refutations*. The advance of science toward the "truth" comes about by discarding theories whose predictions are not confirmed by observations, or theories that are not testable altogether, rather than by shoring up theories with more examples of where they work. *Useful scientific theories are potentially falsifiable.*

Untestable theories are those where a variety of contradictory observations could each be consistent with the theory. For example, consider Freud's psychoanalytic theory. The Oedipus complex theory says that a child is in love with the parent of the opposite sex. A boy desires his mother and wants to destroy his father. If we observe a man to say he loves his mother, that fits in with the theory. If we observe a man to say he hates his mother, that also fits in with the theory, which would say that it is "reaction formation" that leads him to deny his true feelings. In other words, no matter what the man says, it could not falsify the theory because it could be explained by it. Since no observation could potentially falsify the Oedipus theory, its position as a scientific theory could be questioned.

A third, and most reasonable, view is that the progress of science requires both inductive and deductive inference. A particular point of view provides a framework for observations, which lead to a theory that predicts new observations that modify the theory, which then leads to new, predicted observations, and so on toward the elusive "truth," which we generally never reach. Asking which comes first, theory or observation, is like asking which comes first, the chicken or the egg.

In general then, advances in knowledge in the health field come about through constructing, testing, and modifying theories. Epidemiologists make inductive inferences to generalize from many observations, make creative leaps of the imagination to infer explanations and construct theories, and use deductive inferences to test those theories.

Theories, then, can be used to predict observations. But these observations will not always be exactly as we predict them, due to error and the inherent variability of natural phenomena. If the observations are widely different from our predictions we will have to abandon or modify the theory. How do we test the extent of the discordance of our predictions based on theory from the reality of our observations? The test is a statistical or probabilistic test. It is the test of *the null hypothesis, which is the cornerstone of statistical inference* and will be discussed later. Some excellent articles on the logic and philosophy of science, and applications in epidemiology, are listed in the references at the end of this book.[2-6]

1.2   Variability of Phenomena Requires Statistical Analysis

Statistics is a methodology with broad areas of application in science and industry, as well as in medicine and in many other fields. A phenomenon may be principally based on a deterministic model. One example is Boyle's law, which states that for a fixed volume an increase in temperature of a gas determines that there is an increase in pressure. Each time this law is tested the same result occurs. The only variability lies in the error of measurement. Many phenomena in physics and chemistry are of such a nature.

Another type of model is a probabilistic model, which implies that various states of a phenomenon occur with certain probabilities. For instance, the distribution of intelligence is principally probabilistic, that is, given values of intelligence occur with a certain probability in the general population. In biology, psychology, or medicine, where phenomena are influenced by many factors that in themselves are variable and by other factors that are unidentifiable, the models are often probabilistic. In fact, as knowledge in physics has become more refined, it begins to appear that models formerly thought to be deterministic are probabilistic.

In any case, where the model is principally probabilistic, statistical techniques are needed to increase scientific knowledge. *The presence of variation requires the use of statistical analysis.*[7] When there is little variation with respect to a phenomenon, much more weight is given to a small amount of evidence than when there is a great deal of variation. For example, we know that pancreatic cancer appears to be invariably a fatal disease. Thus, if we found a drug that indisputably cured a few patients of pancreatic cancer, we would give a lot of weight to the evidence that the drug represented a cure, far more weight than if the course of this disease were more variable. In contrast to this example, if we were trying to determine whether vitamin C cures colds, we would need to demonstrate its effect in many patients and we would need to use statistical methods to do so, since human beings are quite variable with respect to colds. In fact, in most biological and even more so in social and psychological phenomena, there is a great deal of variability.

1.3   Inductive Inference: Statistics as the Technology of the Scientific Method

Statistical methods are objective methods by which *group trends are abstracted from observations on many separate individuals.* A simple concept of statistics is the calculation of averages, percentages, and so on and the presentation of data in tables and charts. Such techniques for summarizing data are very important indeed and essential to describing the population under study. However, they make up a small part of the field of statistics. A major part of statistics involves the *drawing of inferences from samples to a population* in regard to some characteristic of interest. Suppose we are interested in the average blood pressure of women college students. If we could measure the blood pressure of every single member of this population, we would not have to infer anything. We would simply average all the numbers we obtained. In practice, however, we take a sample of students (properly selected), and on the basis of the data we obtain from the sample, we infer what the mean of the whole population is likely to be.

The reliability of such inferences or conclusions may be evaluated in terms of probability statements. *In statistical reasoning, then, we make inductive inferences, from the particular (sample) to the general (population).* Thus, statistics may be said to be the technology of the scientific method.

1.4   Design of Studies

While the generation of hypotheses may come from anecdotal observations, the testing of those hypotheses must be done by making controlled observations, free of systematic bias. Statistical techniques, to be valid, must be applied to data obtained from well-designed studies. Otherwise, solid knowledge is not advanced.

There are two types of studies: (1) Observational studies, where "Nature" determines who is exposed to the factor of interest and who is not exposed. These studies demonstrate association. Association may imply causation or it may not. (2) Experimental studies, where the investigator determines who is exposed. These may prove causation.

Observational studies may be of three different study designs: *cross-sectional, case-control,* or *prospective*. In a *cross-sectional study* the measurements are taken at one point in time. For example, in a cross-sectional study of high blood pressure and coronary heart disease the investigators determine the blood pressure and the presence of heart disease at the same time. If they find an association, they would not be able to tell which came first. Does heart disease result in high blood pressure or does high blood pressure cause heart disease, or are both high blood pressure and heart disease the result of some other common cause?

In a *case-control study* of smoking and lung cancer, for example, the investigator starts with lung cancer cases and with controls, and through examination of the records or through interviews determines the presence or the absence of the factor in which he or she is interested (smoking). A case-control study is sometimes referred to as a *retrospective study* because data on the factor of interest are collected retrospectively, and thus may be subject to various inaccuracies.

In a *prospective* (or *cohort*) study the investigator starts with a cohort of nondiseased persons with that factor (i.e., those who smoke) and persons without that factor (nonsmokers), and goes forward into some future time to determine the frequency of development of the disease in the two groups. A prospective study is also known as a longitudinal study. *The distinction between case-control studies and prospective studies lies in the sampling. In the case-control study we sample from among the diseased and nondiseased, whereas in a prospective study we sample from among those with the factor and those without the factor.* Prospective studies provide stronger evidence of causality than retrospective studies but are often more difficult, more costly, and sometimes impossible to conduct, for example if the disease under study takes decades to develop or if it is very rare.

In the health field, an experimental study to test an intervention of some sort is called a *clinical trial.* In a clinical trial the investigator assigns patients or participants to one group or another, usually randomly, while trying to keep all other factors constant or controlled for, and compares the outcome of interest in the two (or more) groups. More about clinical trials is in Chapter 6.

In summary, then, the following list is in ascending order of strength in terms of demonstrating causality:

♦ *cross-sectional studies:* useful in showing associations, in providing early clues to etiology.

♦ *case-control studies:* useful for rare diseases or conditions, or when the disease takes a very long time to become manifest (synonymous name: *retrospective studies*).

♦ *cohort studies:* useful for providing stronger evidence of causality, and less subject to biases due to errors of recall or measurement (synonymous names: *prospective studies, longitudinal studies*).

♦ *clinical trials:* prospective, experimental studies that provide the most rigorous evidence of causality.

1.5   How to Quantify Variables

How do we test a hypothesis? First of all, we must set up the hypothesis in a *quantitative* manner. Our criterion measure must be a number of some sort. For example, how many patients died in a drug group compared with how many of the patients died who did not receive the drug, or what is the mean blood pressure of patients on a certain antihypertensive drug compared with the mean blood pressure of patients not on this drug. Sometimes variables are difficult to quantify. For instance, if you are evaluating the quality of care in a clinic in one hospital compared with the clinic of another hospital, it may sometimes be difficult to find a quantitative measure that is representative of quality of care, but nevertheless it can be done and it must be done if one is to test the hypothesis.

There are two types of data that we can deal with: *discrete* or *categorical variables* and *continuous variables.* Continuous variables, theoretically, can assume an infinite number of values between any two fixed points. For example, weight is a continuous variable, as is blood pressure, time, intelligence, and in general, variables in which measurements can be taken. Discrete variables (or categorical variables) are variables that can only assume certain fixed numerical values. For instance, sex is a discrete variable. You may code it as 1 = male, 2 = female, but an individual cannot have a code of 1.5 on sex (at least not theoretically). Discrete variables generally refer to counting, such as the number of patients in a given group who live, the number of people with a certain disease, and so on. In Chapter 3 we will consider a technique for testing a hypothesis where the variable is a discrete one, and subsequently, we will discuss some aspects of continuous variables, but first we will discuss the general concepts of hypothesis testing.

1.6 The Null Hypothesis

The hypothesis we test statistically is called the *null hypothesis.* Let us take a conceptually simple example. Suppose we are testing the efficacy of a new drug on patients with myocardial infarction (heart attack). We divide the patients into two groups—drug and no drug—according to good design procedures, and use as our criterion measure mortality in the two groups. It is our hope that the drug lowers mortality, but to test the hypothesis statistically, we have to set it up in a sort of backward way. We say our hypothesis is that the drug makes no difference, and what we hope to do is to reject the "no difference" hypothesis, based on evidence from our sample of patients. This is known as the *null hypothesis*. We specify our test hypothesis as follows:

H_O (null hypothesis):   death rate in group treated with drug $A =$
                           death rate in group treated with drug B.

This is equivalent to:

H_O: (death rate in group A) – (death rate in group B) $= 0$.

We test this against an *alternate hypothesis,* known as H_A, that the difference in death rates between the two groups *does not* equal 0.

We then gather data and note the *observed* difference in mortality between group A and group B. If this observed difference is sufficiently greater than zero, we reject the null hypothesis. If we reject the null hypothesis of no difference, we accept the *alternate hypothesis*, which is that the drug does make a difference.

When you test a hypothesis, this is the type of reasoning you use:

(1) I will *assume* the hypothesis that there is no difference is true;

(2) I will then collect the data and *observe* the difference between the two groups;

(3) If the null hypothesis is true, how likely is it that *by chance alone* I would get results such as these?

(4) If it is not likely that these results could arise by chance under the assumption than the null hypothesis is true, then I will conclude it is false, and I will "accept" the alternate hypothesis.

1.7   Why Do We Test the Null Hypothesis?

Suppose we believe that drug A is better than drug B in preventing death from a heart attack. Why don't we test that belief directly and see which drug is better, rather than testing the hypothesis that drug A is *equal* to drug B? The reason is that there is an infinite number of ways in which drug A can be better than drug B, so we would have to test an infinite number of hypotheses. If drug A causes 10% fewer deaths than drug B, it is better. So first we would have to see if drug A causes 10% fewer deaths. If it doesn't cause 10% fewer deaths, but if it causes 9% fewer deaths, it is also better. Then we would have to test whether our observations are consistent with a 9% difference in mortality between the two drugs. Then we would have to test whether there is an 8% difference, and so on. Note: each such hypothesis would be set up as a null hypothesis in the following form: Drug A – Drug B mortality = 10%, or equivalently,

(Drug A – Drug B mortality) – (10%) = 0;
(Drug A – Drug B mortality) – (9%) = 0;
(Drug A – Drug B mortality) – (8%) = 0; etc.

On the other hand, when we test the null hypothesis of no difference, we only have to test one value—a 0% difference—and we ask whether our observations are consistent with the hypothesis that there is *no* difference in mortality between the two drugs. If the observations are consistent with a null difference, then we cannot state that one drug is better than the other. If it is unlikely that they are consistent with a null difference, then we can reject that hypothesis and conclude there is a difference.

A common source of confusion arises when the investigator really wishes to show that one treatment is as good as another (in contrast to the above example, where the investigator in her heart of hearts really believes that one drug is better). For example, in the emergency room a quicker procedure may have been devised and the investigator believes it may be as good as the standard procedure, which takes a long time. The temptation in such a situation is to "prove the null hypothesis." *But it is impossible to "prove" the null hypothesis.*

All statistical tests can do is reject the null hypothesis or fail to reject it. We do not prove the hypothesis by gathering affirmative or supportive evidence, because no matter how many times we did the experiment and found a difference close to zero, we could never be assured that the next time we did such an experiment we would not find a huge difference that was nowhere near zero. It is like the example of the white swans discussed earlier: no matter how many white swans we see, we cannot prove that all swans are white, because the next sighting might be a black swan. Rather, we try to falsify or reject our assertion of no difference, and if the assertion of zero difference withstands our attempt at refutation, it survives as a hypothesis in which we continue to have belief. Failure to reject it does not mean we have proven that there is really no difference. It simply means that the evidence we have "is consistent with" the null hypothesis. The results we obtained could have arisen by chance alone if the null hypothesis were true. (Perhaps the design of our study was not appropriate. Perhaps we did not have enough patients.)

So what can one do if one really wants to show that two treatments are equivalent? *One can design a study that is large enough to detect a small difference if there really is one.* If the study has the power (meaning a high likelihood) to detect a difference that is very, very, very small, and one fails to detect it, then one can say with a high degree of confidence that one can't find a meaningful difference between the two treatments. It is impossible to have a study with sufficient power to detect a 0% difference. As the difference one wishes to detect approaches zero, the number of subjects necessary for a given power approaches infinity. The relationships among significance level, power, and sample size are discussed more fully in Chapter 6.

1.8   Types of Errors

The important point is that *we can never be certain* that we are right in either accepting or rejecting a hypothesis. In fact, we run the risk of making one of two kinds of errors: We can reject the null or test hypothesis incorrectly, that is, we can conclude that the drug does reduce mortality when in fact it has no effect. This is known as a *type I error*. Or we can fail to reject the null or test hypothesis incorrectly, that is, we can conclude that the drug does not have an effect when in fact it does reduce mortality. This is known as a *type II error*. Each of these errors carries with it certain consequences. In some cases a type I error may be more serious; in other cases a type II error may be more serious. These points are illustrated in Figure 1.2.

Null Hypothesis (H_O): *Drug has no effect*—no difference in mortality between patients using drug and patients not using drug.

Alternate Hypothesis (H_A): *Drug has effect*—reduces mortality.

		TRUE STATE OF NATURE	
		DRUG HAS NO EFFECT H_O True	DRUG HAS EFFECT; H_O False, H_A True
DECISION ON BASIS OF SAMPLE	DO NOT REJECT H_O No Effect NO	NO ERROR	TYPE II ERROR
	REJECT H_O (Accept H_A) Effect	TYPE I ERROR	NO ERROR

Figure 1.2

If we don't reject H_O, we conclude there is no relationship between drug and mortality. If we do reject H_O and accept H_A, we conclude there is a relationship between drug and mortality.

Actions to be Taken Based on Decision:
(1) If we believe the null hypothesis (i.e., fail to reject it), we will not use the drug.
   Consequences of **wrong** *decision:* Type II error. If we believe H_O incorrectly, since in reality the drug is beneficial, by withholding it we will allow patients to die who might otherwise have lived.

(2) If we reject null hypothesis in favor of the alternate hypothesis, we will use the drug.
   Consequences of **wrong** *decision:* Type I error. If we have rejected the null hypothesis incorrectly, we will use the drug and patients don't benefit. Presuming the drug is not harmful in itself, we do not directly hurt the patients, but since we think we have found the cure, we might no longer test other drugs.

We can never absolutely know the "True State of Nature," but we infer it on the basis of sample evidence.

1.9   Significance Level and Types of Error

We cannot eliminate the risk of making one of these kinds of errors, but we can lower the probabilities that we will make these errors. *The probability of making a type I error is known as the significance level of a statistical test.* When you read in the literature that a result was significant at the .05 level it means that in this experiment the results are such that the probability of making a type I error is less than or equal to .05. Mostly in experiments and surveys people are very concerned about having a low probability of making a type I error and often ignore the type II error. This may be a mistake since in some cases a type II error is a more serious one than a type I error. In designing a study, if you aim to lower the type I error you automatically raise the type II error probability. To lower the probabilities of both the type I and type II error in a study, it is necessary to increase the number of observations.

It is interesting to note that the rules of the Food and Drug Administration (FDA) are set up to lower the probability of making type I errors. In order for a drug to be approved for marketing, the drug company must be able to demonstrate that it does no harm and that it is effective. Thus, many drugs are rejected because their effectiveness cannot be adequately demonstrated. The null hypothesis under test is, "This drug makes no difference." To satisfy FDA rules this hypothesis must be rejected, with the probability of making a type I error (i.e., rejecting it incorrectly) being quite low. In other words, the FDA doesn't want a lot of useless drugs on the market. Drug companies, however, also give weight to guarding against type II error (i.e., avoid believing the no-difference hypothesis incorrectly) so that they may market potentially beneficial drugs.

1.10 Consequences of Type I and Type II Errors

The relative seriousness of these errors depends on the situation. Remember, a type I error (also known as *alpha*) means you are stating something is really there (an effect) when it actually is not, and a type II error (also known as *beta* error) mean you are missing something that is really there. If you are looking for a cure for cancer, a type II error would be quite serious. You would miss finding useful treatments. If you are considering an expensive drug to treat a cold, clearly you would want to avoid a type I error, that is, you would not want to make false claims for a cold remedy.

It is difficult to remember the distinction between type I and II errors. Perhaps this small parable will help us. Once there was a King who was very jealous of his Queen. He had two knights, Alpha, who was very handsome, and Beta, who was very ugly. It happened that the Queen was in love with Beta. The King, however, suspected the Queen was having an affair with Alpha and had him beheaded. Thus, the King made both kinds of errors: he suspected a relationship (with

Alpha) where there was none, and he failed to detect a relationship (with Beta) where there really was one. The Queen fled the kingdom with Beta and lived happily ever after, while the King suffered torments of guilt about his mistaken and fatal rejection of Alpha.

More on alpha, beta, power, and sample size appears in Chapter 6. Since hypothesis testing is based on probabilities, we will first present some basic concepts of probability in Chapter 2.

Chapter 2
A LITTLE BIT OF PROBABILITY

The theory of probability is at bottom nothing but common sense reduced to calculus.

Pierre Simon De Le Place
Theori Analytique des Probabilites (1812–1820)

2.1   What Is Probability?

The probability of the occurrence of an event is indicated by a number ranging from 0 to 1. An event whose probability of occurrence is 0 is certain not to occur, whereas an event whose probability is 1 is certain to occur.

The classical definition of probability is as follows: if an event can occur in N mutually exclusive, equally likely ways and if n_A of these outcomes have attribute A, then the probability of A, written as $P(A)$, equals n_A/N. This is an a priori definition of probability, that is, one determines the probability of an event before it has happened. Assume one were to toss a die and wanted to know the probability of obtaining a number divisible by three on the toss of a die. There are six possible ways that the die can land. Of these, there are two ways in which the number on the face of the die is divisible by three, a 3 and a 6. Thus, the probability of obtaining a number divisible by three on the toss of a die is 2/6 or 1/3.

In many cases, however, we are not able to enumerate all the possible ways in which an event can occur, and, therefore, we use the *relative frequency definition of probability*. This is defined as the number of times that the event of interest has occurred divided by the total number of trials (or opportunities for the event to occur). Since it is based on previous data, it is called the *a posteriori definition of probability*.

For instance, if you select at random a white American female, the probability of her dying of heart disease is .00287. This is based on the

19

finding that per 100,000 white American females, 287 died of coronary heart disease (estimates are for 2001, National Center for Health Statistics[7]). When you consider the probability of a white American female who is between ages 45 and 64, the figure drops to .00088 (or 88 women in that age group out of 100,000), and when you consider women 65 years and older, the figure rises to .01672 (or 1672 per 100,000). For white men 65 or older it is .0919 (or 9190 per 100,000). The two important points are (1) to determine a probability, *you must specify the population to which you refer*, for example, all white females, white males between 65 and 74, nonwhite females between 65 and 74, and so on; and (2) the *probability figures are constantly revised* as new data become available.

This brings us to the notion of *expected frequency*. If the probability of an event is P and there are N trials (or opportunities for the event to occur), then we can expect that the event *will* occur $N \times P$ times. It is necessary to remember that probability "works" for large numbers. When in tossing a coin we say the probability of it landing on heads is .50, we mean that in many tosses half the time the coin will land heads. If we toss the coin ten times, we may get three heads (30%) or six heads (60%), which are a considerable departure from the 50% we expect. But if we toss the coin 200,000 times, we are very likely to be close to getting exactly 100,000 heads or 50%.

Expected frequency is really the way in which probability "works." It is difficult to conceptualize applying probability to an individual. For example, when TV announcers proclaim there will be say, 400 fatal accidents in State X on the Fourth of July, it is impossible to say whether any individual person will in fact have such an accident, but we can be pretty certain that the number of such accidents will be very close to the predicted 400 (based on probabilities derived from previous Fourth of July statistics).

2.2   Combining Probabilities

There are two laws for combining probabilities that are important. First, if there are *mutually exclusive events* (i.e., if one occurs, the other cannot), the probability of either one or the other occurring is the *sum* of their individual probabilities. Symbolically,

$$P(A \text{ or } B) = P(A) + P(B)$$

An example of this is as follows: the probability of getting either a 3 or a 4 on the toss of a die is 1/6 + 1/6 = 2/6.

A useful thing to know is that the sum of the individual probabilities of all possible mutually exclusive events must equal 1. For example, if A is the event of winning a lottery, and not A (written as $\overline{A}$), is the event of not winning the lottery, then $P(A) + P(\overline{A}) = 1.0$ and $P(\overline{A}) = 1 - P(A)$.

Second, if there are two independent events (i.e., the occurrence of one is not related to the occurrence of the other), the joint probability of their occurring together (jointly) is the *product* of the individual probabilities. Symbolically,

$$P(A \text{ and } B) = P(A) \times P(B)$$

An example of this is the probability that on the toss of a die you will get a number that is both even and divisible by 3. This probability is equal to $1/2 \times 1/3 = 1/6$. (The only number both even and divisible by 3 is the number 6.)

The joint probability law is used to test whether events are independent. If they are independent, the product of their individual probabilities should equal the joint probability. If it does not, they are not independent. It is the basis of the chi-square test of significance, which we will consider in the next section.

Let us apply these concepts to a medical example. The mortality rate for those with a heart attack in a special coronary care unit in a certain hospital is 15%. Thus, the probability that a patient with a heart attack admitted to this coronary care unit will die is .15 and that he will survive is .85. If two men are admitted to the coronary care unit on a particular day, let A be the event that the first man dies and let B be the event that the second man dies.

The probability that both will die is

$$P(A \text{ and } B) = P(A) \times P(B) = .15 \times .15 = .0225$$

We assume these events are independent of each other so we can multiply their probabilities. Note, however, that the probability that either one *or* the other will die from the heart attack is *not* the sum of their probabilities because these two events are not mutually exclusive. It is possible that both will die (i.e., both A and B can occur).

To make this clearer, a good way to approach probability is through the use of Venn diagrams, as shown in Figure 2.1. Venn diagrams consist of squares that represent the universe of possibilities and circles that define the events of interest.

In diagrams 1, 2, and 3, the space inside the square represents all N possible outcomes. The circle marked A represents all the outcomes that constitute event A; the circle marked B represents all the outcomes that constitute event B. Diagram 1 illustrates two mutually exclusive events; an outcome in circle A cannot also be in circle B. Diagram 2 illustrates two events that can occur jointly: an outcome in circle A can also be an outcome belonging to circle B. The shaded area marked AB represents outcomes that are the occurrence of both A and B. The diagram 3 represents two events where one (B) is a subset of the other (A); an outcome in circle B must also be an outcome constituting event A, but the reverse is not necessarily true.

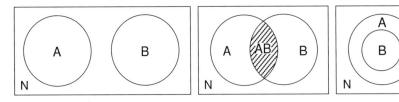

Figure 2.1

It can be seen from diagram 2 that if we want the probability of an outcome being either A *or* B and if we add the outcomes in circle A to the outcomes in circle B, we have added in the outcomes in the shaded area twice. Therefore, we must subtract the outcomes in the shaded area (*A and B*) also written as (*AB*) once to arrive at the correct answer. Thus, we get the result

$$P(A \text{ or } B) = P(A) + P(B) - P(AB)$$

2.3   Conditional Probability

Now let us consider the case where the chance that a particular event happens is dependent on the outcome of another event. The probability of *A*, given that *B* has occurred, is called the conditional probability of *A* given *B*, and is written symbolically as $P(A|B)$. An illustration of this is provided by Venn diagram 2. When we speak of conditional probability, the denominator becomes all the outcomes in circle B (instead of all *N* possible outcomes) and the numerator consists of those outcomes that are in that part of *A* which also contains outcomes belonging to *B*. This is the shaded area in the diagram labeled AB. If we return to our original definition of probability, we see that

$$P(A \mid B) = \frac{n_{AB}}{n_B}$$

(the number of outcomes in both *A and B*, divided by the total number of outcomes in *B*).

If we divide both numerator and denominator by *N*, the total number of *all* possible outcomes, we obtain

$$P(A \mid B) = \frac{n_{AB}/N}{n_B/N} = \frac{P(A \text{ and } B)}{P(B)}$$

Multiplying both sides by $P(B)$ gives the *complete* multiplicative law:

$$P(A \text{ and } B) = P(A \mid B) \times P(B)$$

Of course, if A and B are independent, then the probability of A given B is just equal to the probability of A (since the occurrence of B does not influence the occurrence of A) and we then see that

$$P(A \text{ and } B) = P(A) \times P(B)$$

2.4   Bayesian Probability

Imagine that M is the event "loss of memory," and B is the event "brain tumor." We can establish from research on brain tumor patients the probability of *memory loss given a brain tumor, P(M|B).* A clinician, however, is more interested in the probability of *a brain tumor, given that a patient has memory loss, P(B | M).*

It is difficult to obtain that probability directly because one would have to study the vast number of persons with memory loss (which in most cases comes from other causes) and determine what proportion of them have brain tumors.

Bayes' equation (or Bayes' theorem) estimates $P(B | M)$ as follows:

$$P(\text{brain tumor, given memory loss}) = \frac{P(\text{memory loss, given brain tumor}) \times P(\text{brain tumor})}{P(\text{memory loss})}$$

In the denominator, the event of "memory loss" can occur either among people with brain tumor, with probability $= P(M | B) \, P(B)$, or among people with no brain tumor, with probability $= P(M | \bar{B}) P(\bar{B})$. Thus,

$$P(B | M) = \frac{P(M | B) P(B)}{P(M | B) P(B) + P(M | \bar{B}) P(\bar{B})}$$

The overall probability of a brain tumor, $P(B)$ is the "a priori probability," which is a sort of "best guess" of the prevalence of brain tumors.

2.5 Odds and Probability

When the odds of a particular horse *losing* a race are said to be 4 to 1, he has a 4/5 = .80 probability of losing. To convert an odds statement to probability, we add 4 + 1 to get our denominator of 5. The odds of the horse *winning* are 1 to 4, which means he has a probability of winning of 1/5 = .20.

$$\text{The odds in favor of } A = \frac{P(A)}{P(not\ A)} = \frac{P(A)}{1 - P(A)}$$

$$P(A) = \frac{odds}{1\ +\ odds}$$

The odds of drawing an ace = 4 (aces in a deck) to 48 (cards that are not aces) = 1 to 12; therefore, $P(ace) = 1/13$. The odds *against* drawing an ace = 12 to 1; $P(Not\ Ace) = 12/13$.

In medicine, odds are often used to calculate an *odds ratio*. An odds ratio is simply the ratio of two odds. For example, say that in a particular study comparing lung cancer patients with controls, it was found that the odds of being a lung cancer case for people who smoke were 5 to 4 (5/4) and the odds of having lung cancer for nonsmokers was 1 to 8 (1/8), then the odds ratio would be

$$\frac{5/4}{1/8}\ =\ \frac{5 \times 8}{4 \times 1}\ =\ \frac{40}{4}\ =\ 10$$

An odds ratio of 10 means that the odds of being a lung cancer case is 10 times greater for smokers than for nonsmokers.

Note, however, that we cannot determine from such an analysis what the probability of getting lung cancer is for smokers, because in order to do that we would have to know how many people out of all smokers developed lung cancer, and we haven't studied all smokers; all we do know is how many out of all our lung cancer cases were smokers. Nor can we get the probability of lung cancer among nonsmokers, because we would have to a look at a population of nonsmokers and see how many of them developed lung cancer. All we do know is that smokers have 10-fold greater odds of having lung cancer than non-smokers.

More on this topic is presented in Section 4.12.

2.6   Likelihood Ratio

A related concept is the likelihood ratio (LR), which tells us how likely it is that a certain result would arise from one set of circumstances in relation to how likely the result would arise from an opposite set of circumstances.

For example, if a patient has a sudden loss of memory, we might want to know the likelihood ratio of that symptom for a brain tumor, say. What we want to know is the likelihood that the memory loss arose out of the brain tumor *in relation to* the likelihood that it arose from some other condition. The likelihood ratio is a ratio of conditional probabilities.

$$LR = \frac{P(memory\ loss,\ given\ brain\ tumor)}{P(memory\ loss,\ given\ no\ brain\ tumor)}$$

$$= \frac{P(M\ given\ B)}{P(M\ given\ not\ B)}$$

Of course to calculate this LR we would need to have estimates of the probabilities involved in the equation, that is, we would need to know the following: among persons who have brain tumors, what proportion have memory loss, and among persons who don't have brain tumors, what proportion have memory loss. It may sometimes be quite difficult to establish the denominator of the likelihood ratio because we would need to know the prevalence of memory loss in the general population.

The LR is perhaps more practical to use than the Bayes' theorem, which gives the probability of a particular disease given a particular symptom. In any case, it is widely used in variety of situations because it addresses this important question: If a patient presents with a symptom, what is the likelihood that the symptom is due to a particular disease *rather than* to some other reason than this disease?

2.7 Summary of Probability

Additive Law:

$$P(A \ or \ B) = P(A) + P(B) - P(A \ and \ B)$$

If events are mutually exclusive: $P(A \ or \ B) = P(A) + P(B)$.

Multiplicative Law:

$$P(A \ and \ B) = P(A \mid B) \times P(B)$$

If events are independent: $P(A \ and \ B) = P(A) \times P(B)$.

Conditional Probability:

$$P(A \mid B) = \frac{P(A \ and \ B)}{P(B)}$$

Bayes' Theorem:

$$P(B \mid A) = \frac{P(A \mid B) \, P(B)}{P(A \mid B) \, P(B) + P(A \mid \bar{B}) \, P(\bar{B})}$$

Odds of A:

$$\frac{P(A)}{1 \ - \ P(A)}$$

Likelihood Ratio:

$$\frac{P(A \mid B)}{P(A \mid \overline{B})}$$

Chapter 3
MOSTLY ABOUT STATISTICS

A statistician is someone who, with his head in an oven and his feet in a bucket of ice water, when asked how he feels, responds: "On the average, I feel fine."

Anonymous

Different statistical techniques are appropriate depending on whether the variables of interest are discrete or continuous. We will first consider the case of discrete variables and present the chi-square test and then we will discuss methods applicable to continuous variables.

3.1   Chi-Square for 2 × 2 Tables

The chi-square test is a statistical method to determine whether the results of an experiment may arise by chance or not. Let us, therefore, consider the example of testing an anticoagulant drug on female patients with myocardial infarction. We hope the drug lowers mortality, but we set up our null hypothesis as follows:

- ◆ Null Hypothesis   There is no difference in mortality between the treated group of patients and the control group.

- ◆ Alternate Hypothesis: The mortality in the treated group is lower than in the control group.

(The data for our example come from a study done a long time ago and refer to a specific high-risk group.[8] They are used for illustrative purposes and they do not reflect current mortality rates for people with myocardial infarction.)

We then record our data in a 2 × 2 *contingency* table in which each patient is classified as belonging to one of the four cells:

Observed Frequencies

	Control	Treated	
Lived	89	223	312
Died	40	39	79
Total	129	262	391

The mortality in the control group is 40/129 = 31% and in the treated it is 39/262 = 15%. But could this difference have arisen by chance? We use the chi-square test to answer this question. What we are really asking is whether the two categories of classification (control vs. treated by lived vs. died) are independent of each other. If they are independent, what frequencies would we expect in each of the cells? And *how different are our observed frequencies from the expected ones?* How do we measure the size of the difference?

To determine the expected frequencies, consider the following:

	Control	Treated	
Lived	a	b	$(a + b)$
Died	c	d	$(c + d)$
Total	$(a + c)$	$(b + d)$	N

If the categories are independent, then the probability of a patient being both a control and living is P(control) × P(lived). [Here we apply the law referred to in Chapter 2 on the joint probability of two independent events.]

The expected frequency of an event is equal to the probability of the event times the number of trials = N × P. So the expected number of patients who are *both* controls *and* live is

$$N \times P(control\ and\ lived) = N \times P(control) \times P(lived)$$

$$= N \left[\frac{(a + c)}{N} \times \frac{(a + b)}{N} \right] = (a + c) \times \frac{(a + b)}{N}$$

In our case this yields the following table:

	Control	Treated	
Lived	$129 \times \dfrac{312}{391} = 103$	$262 \times \dfrac{312}{391} = 209$	312
Died	$129 \times \dfrac{79}{391} = 26$	$262 \times \dfrac{79}{391} = 53$	79
Total	129	262	391

Another way of looking at this is to say that since 80% of the patients in the total study lived (i.e., 312/391 = 80%), we would expect that
80% of the control patients and 80% of the treated patients would live.
These expectations differ, as we see, from the observed frequencies
noted earlier, that is, those patients treated did, in fact, have a lower
mortality than those in the control group.

Well, now that we have a table of observed frequencies and a table
of expected values, how do we know just how different they are? Do
they differ just by chance or is there some other factor that causes
them to differ? To determine this, we calculate a value called
chi-square (also written as χ^2). This is obtained by taking the observed
value in each cell, subtracting from it the expected value in each cell,
squaring this difference, and dividing by the expected value for each
cell. When this is done for each cell, the four resulting quantities are
added together to give a number called chi-square. Symbolically this
formula is as follows:

$$\frac{(O_a-e_a)^2}{e_a} + \frac{(O_b-e_b)^2}{e_b} + \frac{(O_c-e_c)^2}{e_c} + \frac{(O_d-e_d)^2}{e_d}$$

where O is the observed frequency and e is the expected frequency in each cell.

This number, called chi-square, is a statistic that has a known distribution. What that means, in essence, is that for an infinite number of such 2×2 tables, chi-squares have been calculated and we thus know what the probability is of getting certain values of chi-square. Thus, when we calculate a chi-square for a particular 2×2 contingency table, we know how likely it is that we could have obtained a value as large as the one that we actually obtained strictly by chance, under the assumption the hypothesis of independence is the correct one, that is, if the two categories of classification were unrelated to one another or if the null hypothesis were true. The particular value of chi-square that we get for our example happens to be 13.94.

From our knowledge of the distribution of values of chi-square, we know that if our null hypothesis is true, that is, if there is no difference in mortality between the control and treated group, then the probability that we get a value of chi-square as large or larger than 13.94 by chance alone is very, very low; in fact this probability is less than .005. Since it is not likely that we would get such a large value of chi-square by chance under the assumption of our null hypothesis, *it must be that it has arisen not by chance but because our null hypothesis is incorrect*. We, therefore, reject the null hypothesis at the .005 level of significance and accept the alternate hypothesis, that is, we conclude that among women with myocardial infarction the new drug does reduce mortality. The probability of obtaining these results by chance alone is less than 5/1000 (.005). Therefore, the probability of rejecting the null hypothesis, when it is in fact true (type I error) is less than .005.

The probabilities for obtaining various values of chi-square are tabled in most standard statistics texts, so that the procedure is to calculate the value of chi-square and then look it up in the table to determine whether or not it is significant. That value of chi-square that must be obtained from the data in order to be significant is called the *critical value.* The critical value of chi-square at the .05 level of significance for a 2 × 2 table is 3.84. This means that when we get a value of 3.84 *or greater* from a 2 × 2 table, we can say there is a significant difference between the two groups. Appendix A provides some critical values for chi-square and for other tests.

In actual usage, a correction is applied for 2 × 2 tables known as the Yates' correction and calculation is done using the formula:

$$\frac{N \left[\mid ad - bc \mid - \dfrac{N}{2} \right]^2}{(a + b)(c + d)(a + c)(b + d)}$$

Note: $\mid ad - bc \mid$ means the absolute value of the difference between $a \times d$ and $b \times c$. In other words, if $a \times d$ is greater than $b \times c$, subtract bc from ad; if bc is greater than ad, subtract ad from bc. The corrected chi-square so calculated is 12.95, still well above the 3.84 required for significance.

The chi-square test should not be used if the numbers in the cells are too small. The rules of thumb: When the total N is greater than 40, use the chi-square test with Yates' correction. When N is between 20 and 40 and the expected frequency in each of the four cells is 5 or more, use the corrected chi-square test. If the smallest expected frequency is less than 5, or if N is less than 20, use the Fisher's test.

While the chi-square test approximates the probability, the Fisher's Exact Test gives the exact probability of getting a table with values like those obtained or even more extreme. A sample calculation is shown in Appendix B. The calculations are unwieldy but the Fisher's exact test is also usually included in most statistics programs for personal computers. More on this topic may be found in the book *Statistical Methods for Rates and Proportions* by Joseph L. Fleiss. The important thing is to know when the chi-square test is or is not appropriate.

3.2 McNemar Test

Suppose we have the situation where measurements are made on the same group of people before and after some intervention, or suppose we are interested in the agreement between two judges who evaluate the same group of patients on some characteristics. In such situations, the before and after measures, or the opinions of two judges, are not independent of each other, since they pertain to the same individuals. Therefore, the Chi-Square test or the Fisher's Exact Test are not appropriate. Instead, we can use the McNemar test.

Consider the following example. Case histories of patients who were suspected of having ischemic heart disease (a decreased blood flow to the heart because of clogging of the arteries), were presented to two cardiology experts. The doctors were asked to render an opinion on the basis of the available information about the patient. They could recommend either (1) that the patient should be on medical therapy or (2) that the patient have an angiogram, which is an invasive test, to determine if the patient is a suitable candidate for coronary artery bypass graft surgery (known as CABG). Table 3.1 shows the results of these judgments on 661 patients.

TABLE 3.1

		EXPERT 1 Medical	EXPERT 1 Surgical	
EXPERT 2	Medical	$a = 397$	$b = 97$	$a + b = 494$
EXPERT 2	Surgical	$c = 91$	$d = 76$	$c + d = 167$
		$a + c = 488$	$b + d = 173$	$N = 661$

Note that in cell b Expert 1 advised surgery and Expert 2 advised medical therapy for 97 patients, whereas in cell c Expert 1 advised medical therapy and Expert 2 advised surgery for 91 of the patients. Thus, the two physicians disagree in 188 of the 661 cases or 28% of the time. Cells a and d represent patients about whom the two doctors agree. They agree in 473 out the 661 case or 72% of the time.

To determine whether the observed disagreement could have arisen by chance alone under the null hypothesis of no real disagreement in recommendations between the two experts, we calculate a type of chi-square value as follows:

$$\chi^2 \ (chi\text{-}square) = \frac{(\,|\,b - c\,| - 1)^2}{(b + c)} = \frac{25}{188} = .13$$

($\,|\,b - c\,|$ means the absolute value of the difference between the two cells, that is, irrespective of the sign; the -1 in the numerator is analogous to the Yates' correction for chi-square described in Section 3.1, and gives a better approximation to the chi-square distribution.) A chi-square of .13 does not reach the critical value of chi-square of 3.84 needed for a .05 significance level, as described in Section 3.1, so we cannot reject the null hypothesis and we conclude that our data are consistent with no difference in the opinions of the two experts. Were the chi-square test significant, we would have to reject the null hypothesis and say the experts significantly disagree. However, such a test does not tell us about the *strength* of their agreement, which can be evaluated by a statistic called Kappa.

3.3 Kappa

The two experts could be agreeing just by chance alone, since both experts are more likely to recommend medical therapy for these patients. Kappa is a statistic that tells us the extent of the agreement between the two experts above and beyond chance agreement.

$$K = \frac{\textit{Proportion of observed agreement} - \textit{Proportion of agreement by chance}}{1 - \textit{Proportion of agreement by chance}}$$

To calculate the expected number of cases in each cell of the table, we follow the procedure described for chi-square in Section 3.1. The cells a and d in Table 3.1 represent agreement. The expected number by chance alone is

$$cell\ a: \quad \frac{494 \times 488}{661} = 365$$

$$cell\ d: \quad \frac{167 \times 173}{661} = 44$$

So the proportion of agreement expected by chance alone is

$$\frac{365 + 44}{661} = .619$$

that is, by chance alone the experts would be expected to agree 62% of the time. The proportion of observed agreement is

$$\frac{397 + 76}{661} = .716$$

$$Kappa = \frac{.716 - .619}{1 - .619} = \frac{.097}{.381} = .25$$

If the two experts agreed at the level of chance only, Kappa would be 0; if the two experts agreed perfectly Kappa would be 1. The topic of Kappa is thoroughly described in the book by Fleiss listed in the Suggested Readings.

3.4   Description of a Population: Use of the Standard Deviation

In the case of continuous variables, as for discrete variables, we may be interested in description or in inference. When we wish to describe a population with regard to some characteristic, we generally use the mean or average as an index of *central tendency* of the data.

Other measures of central tendency are the *median* and the *mode*. The median is that value above which 50% of the other values lie and below which 50% of the values lie. It is the middle value or the 50th percentile. To find the median of a set of scores we arrange them in ascending (or descending) order and locate the middle value if there are an odd number of scores, or the average between the two middle scores if there are an even number of scores. The mode is the value that occurs with the greatest frequency. There may be several modes in a set of scores but only one median and one mean value. These definitions are illustrated below. The mean is the measure of central tendency most often used in inferential statistics.

Measures of Central Tendency	
Set of scores	Ordered
12	6
12	8
6	10
8	**11 Median**
11	*12 Mode*
10	*12*
15	15
SUM: 74	*Mean* = 74/7 = 10.6

The true mean of the population is called m and we estimate that mean from data obtained from a sample of the population. The sample mean is called $\bar{x}$ (read as x bar). We must be careful to specify exactly the population from which we take a sample. For instance, in the general population the average I.Q. is 100, but the average I.Q. of the population of children age 6 to 11 years whose fathers are college graduates is 112.[9] Therefore, if we take a sample from either of these populations, we would be estimating a different population mean and we must specify to which population we are making inferences.

However, the mean does not provide an adequate description of a population. What is also needed is some measure of *variability* of the data around the mean. Two groups can have the same mean but be very different. For instance, consider a hypothetical group of children each of whose individual I.Q. is 100; thus, the mean is 100. Compare this to another group whose mean is also 100 but includes individuals with I.Q.'s is of 60 and those with I.Q.s of 140. Different statements must be made about these two groups: one is composed of all average individuals; the other includes both retardates and geniuses.

The most commonly used index of variability is the *standard devia-tion (s.d.)*, which is a type of measure related to the average distance of the scores from their mean value. The square of the standard deviation is called *variance*. The population standard deviation is denoted by the Greek letter σ (sigma). When it is calculated from a *sample*, it is written as s.d. and is illustrated in the example below:

I.Q. scores		Deviations from mean		Squared scores for B
Group A	Group B	$x_i - \overline{x}_B$	$(x_i - \overline{x}_B)^2$	x_B^2
100	60	−40	1600	3600
100	140	40	1600	19,600
100	80	−20	400	6400
100	120	20	400	14,400
$\Sigma = 400$	$\Sigma = 400$	$\Sigma = 0$	$\Sigma = 4000$ of squared deviations	$\Sigma = 44,000$ sum of squares
$\overline{x}_A$ = mean	$\overline{x}_B$ = mean			
= 100	= 100			

Note: The symbol "Σ" means "sum."

Note: The sum of deviations from the mean, as in column 3, is always 0; that is why we sum the squared deviations, as in column 4.

$$\overline{x}_A = mean = \frac{400}{4} = 100; \qquad \overline{x}_B = \frac{400}{4} = 100$$

$$s.d. = \sqrt{\frac{\sum of\ (each\ value\ -\ mean\ of\ group)^2}{n-1}} = \sqrt{\frac{\sum(x_i - \overline{x})^2}{n-1}}$$

$$s.d._A = \frac{0}{3} = 0;$$

(In Group A since each score is equal to the mean of 100, there are no deviations from the mean of A.)

$$s.d._B = \sqrt{\frac{4000}{3}} = \sqrt{1333} = 36.51$$

An equivalent formula for s.d. that is more suited for actual calculations is

$$s.d. = \sqrt{\frac{\sum x_i^2 - n\overline{x}^2}{n-1}}$$

For group B we have

$$s.d. = \sqrt{\frac{44000 - 4(100)^2}{3}} = \sqrt{\frac{44000 - 40000}{3}} = \sqrt{\frac{4000}{3}} = 36.51$$

$Variance = (s.d.)^2$

Note the mean of both groups is 100 but the standard deviation of group A is 0 while the s.d. of group B is 36.51. (We divide the squared deviations by $n - 1$, rather than by n because we are estimating the population σ from sample data, and dividing by $n - 1$ gives a better estimate. The mathematical reason is complex and beyond the scope of this book.)

3.5   Meaning of the Standard Deviation: The Normal Distribution

The standard deviation is a measure of the dispersion or spread of the data. Consider a variable like I.Q., which is normally distributed, that is, it can be described by the familiar, bell-shaped curve where most of the values fall around the mean with decreasing number of values at either extremes. In such a case, 68% of the values lie within 1 standard deviation on either side of the mean, 95% of the values lie within 2 standard deviations of the mean, and 99% of the values lie within 3 standard deviations of the mean.

This is illustrated in Figure 3.1.

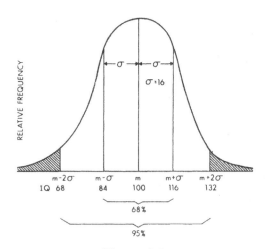

Figure 3.1

In the population at large, 95% of people have I.Q.s between 68 and 132. Approximately 2.5% of people have I.Q.s above 132 and another 2.5% of people have I.Q.s below 68. (This is indicated by the shaded areas at the tails of the curves.)

If we are estimating from a sample and if there are a large number of observations, the standard deviation can be estimated from the *range of the data,* that is, the difference between the smallest and the highest value. Dividing the range by 6 provides a rough estimate of the standard deviation if the distribution is normal, because 6 standard deviations (3 on either side of the mean) encompass 99%, or virtually all, of the data.

On an individual, clinical level, knowledge of the standard deviation is very useful in deciding whether a laboratory finding is normal, in the sense of "healthy." Generally a value that is more than 2 standard deviations away from the mean is suspect, and perhaps further tests need to be carried out.

For instance, suppose as a physician you are faced with an adult male who has a hematocrit reading of 39. Hematocrit is a measure of the amount of packed red cells in a measured amount of blood. A low hematocrit may imply anemia, which in turn may imply a more serious condition. You also know that the average hematocrit reading for adult males is 47. Do you know whether the patient with a reading of 39 is normal (in the sense of healthy) or abnormal? You need to know the standard deviation of the distribution of hematocrits in people before you can determine whether 39 is a normal value. In point of fact, the standard deviation is approximately 3.5; thus, plus or minus 2 standard deviations around the mean results in the range of from 40 to 54 so that 39 would be slightly low. For adult females, the mean hematocrit is 42 with a standard deviation of 2.5, so that the range of plus or minus 2 standard deviations away from the mean is from 37 to 47. Thus, if an adult female came to you with a hematocrit reading of 39, she would be considered in the "normal" range.

3.6   The Difference Between Standard Deviation and Standard Error

Often data in the literature are reported as ± s.d. (read as mean + or −1 standard deviation). Other times they are reported as ± s.e. (read as mean + or −1 standard error). *Standard error* and *standard deviation* are often confused, but they serve quite different functions. To understand the concept of standard error, you must remember that the purpose of statistics is to draw inferences from samples of data to the population from which these samples came. Specifically, we are interested in estimating the true mean of a population for which we have a sample mean based on, say, 25 cases. Imagine the following:

Population I.Q. scores, x_i	Sample means based on 25 people randomly selected
110	$\overline{x_1} = 102$
100	
105	$\overline{x_2} = 99$
98	
140	$\overline{x_3} = 101$
—	$\overline{x_4} = 98$
—	—
100	100

m = mean of all the x_i's

$m_{\overline{x}} = m$
mean of the means is m, the population mean

σ = population standard deviation

$\dfrac{\sigma}{\sqrt{n}} =$ standard deviation of the distribution of the $\overline{x}$'s called the standard error of the mean = $\sigma_{\overline{x}}$

There is a population of I.Q. scores whose mean is 100 and its standard deviation is 16. Now imagine that we draw a sample of 25 people at random from that population and calculate the sample mean $\bar{x}$. This sample mean happens to be 102. If we took another sample of 25 individuals we would probably get a slightly different sample mean, for example 99. Suppose we did this repeatedly an infinite (or a very large) number of times, each time throwing the sample we just drew back into the population pool from which we would sample 25 people again. We would then have a very large number of such sample means. These sample means would form a normal distribution. Some of them would be very close to the true population mean of 100, and some would be at either end of this "distribution of means" as in Figure 3.2.

This distribution of sample means would have its own standard deviation, that is, a measure of the spread of the data around the mean of the data. In this case, the data are sample means rather than individual values. The standard deviation of this distribution of means is called the *standard error of the mean*.

It should be pointed out that this distribution of means, which is also called the sampling distribution of means, is a theoretical construct. Obviously, we don't go around measuring samples of the population to construct such a distribution. Usually, in fact, we just take *one sample of 25* people and imagine what this distribution might be. However, due to certain mathematical derivations, we know a lot about this theoretical distribution of population means and therefore we can draw important inferences based on just one sample mean. What we do know is that the distribution of means is a normal distribution, that its mean is the same as the population mean of the individual values, that is, *the mean of the means is m*, and that its standard deviation is equal to the standard deviation of the original individual values divided by the square root of the number of people in the sample.

Standard error of the mean =

$$\sigma_{\bar{x}} = \frac{\sigma}{\sqrt{n}}$$

In this case it would be

$$\frac{16}{\sqrt{25}} = \frac{16}{5} = 3.2$$

The distribution of means would look as shown in Figure 3.2.

Please note that when we talk about population values, which we usually don't know but are trying to estimate, we refer to the mean as m and the standard deviation as σ. When we talk about values calculated from samples, we refer to the mean as $\bar{x}$, the standard deviation as s.d., and the standard error as s.e.

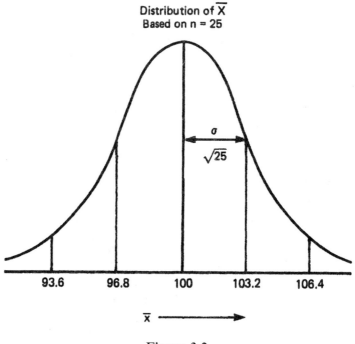

Figure 3.2

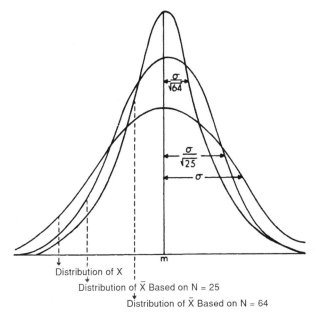

Figure 3.3

Now imagine that we have a distribution of means based on samples of 64 individuals. The mean of these means is also *m*, but its dispersion, or standard error, is smaller. It is $16/\sqrt{64}=16/8=2$. This is illustrated in Figure 3.3.

It is easily seen that if we take a sample of 25 individuals, their mean is likely to be closer to the true mean than the value of a single individual, and if we draw a sample of 64 individuals, their mean is likely to be even closer to the true mean than was the mean we obtained from the sample of 25. Thus, the larger the sample size, the better is our estimate of the true population mean.

The standard deviation is used to describe the dispersion or variability of the scores. The standard error is used to draw inferences about the population mean from which we have a sample. We draw such inferences by constructing confidence intervals, which are discussed in Section 3.11.

3.7   Standard Error of the Difference Between Two Means

This concept is analogous to the concept of standard error of the mean. The standard error of the differences between two means is the standard deviation of a theoretical distribution of differences between two means. Imagine a group of men and a group of women each of whom have an I.Q. measurement. Suppose we take a sample of 64 men and a sample of 64 women, calculate the mean I.Q.s of these two samples, and obtain their differences. If we were to do this an infinite number of times, we would get a *distribution of differences* between sample means of two groups of 64 each. These difference scores would be normally distributed; their mean would be the true average difference between the populations of men and women (which we are trying to infer from the samples), and the standard deviation of this distribution is called the *standard error of the differences between two means.*

The standard error of the difference between two means of populations X and Y is given by the formula

$$\sigma_{\bar{x}-\bar{y}} = \sqrt{\frac{\sigma_x^2}{n_x} + \frac{\sigma_y^2}{n_y}}$$

where σ_x^2 is the variance of population X and σ_y^2 is the variance of population Y; n_x is the number of cases in the sample from population X and n_y is the number of cases in the sample from population Y.

In some cases we know or assume that the variances of the two populations are equal to each other and that the variances that we calculate from the samples we have drawn are both estimates of a common variance. In such a situation, we would want to pool these estimates to get a better estimate of the common variance. We denote this *pooled estimate* as $s_{pooled}^2 = s_p^2$ and we calculate the standard error of the difference between means as

$$s.e._{\bar{x}-\bar{y}} = \sqrt{s_p^2\left(\frac{1}{n_x} + \frac{1}{n_y}\right)} = s_p\sqrt{\frac{1}{n_x} + \frac{1}{n_y}}$$

We calculate s_p^2 from sample data:

$$s_p^2 = \frac{(n_x - 1)s_x^2 + (n_y - 1)s_y^2}{n_x + n_y - 2}$$

This is the equivalent to

$$s_p^2 = \frac{\Sigma(x_i - \bar{x})^2 + \Sigma(y_i - \bar{y})^2}{n_x + n_y - 2}$$

Since in practice we will always be calculating our values from sample data, we will henceforth use the symbology appropriate to that.

3.8 Z Scores and the Standardized Normal Distribution

The standardized normal distribution is one whose mean = 0, standard deviation = 1, and the total area under the curve = 1. The standard normal distribution looks like Figure 3.4.

On the abscissa, instead of x we have a transformation of x called the standard score, Z. Z is derived from x by the following:

$$Z = \frac{x - m}{\sigma}$$

Thus, the Z score really tells you how many standard deviations from the mean a particular x score is.

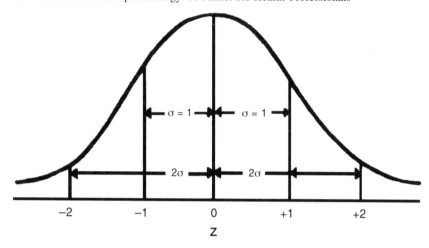

Figure 3.4

Any distribution of a normal variable can be transformed to a distribution of Z by taking each x value, subtracting from it the mean of x (i.e., m), and dividing this deviation of x from its mean, by the standard deviation. Let us look at the I.Q. distribution again in Figure 3.5.

Thus, an I.Q. score of 131 is equivalent to a Z score of 1.96 (i.e., it is 1.96, or nearly 2, standard deviations above the mean I.Q.).

$$Z = \frac{131 - 100}{16} = 1.96$$

One of the nice things about the Z distribution is that the probability of a value being anywhere between two points is equal to the area under the curve between those two points. (Accept this on faith.) It happens that the area to the right of 1.96 corresponds to a probability of .025, or 2.5% of the total curve. Since the curve is symmetrical, the probability of Z being to the left of −1.96 is also .025. Invoking the ad-

DISTRIBUTION OF X

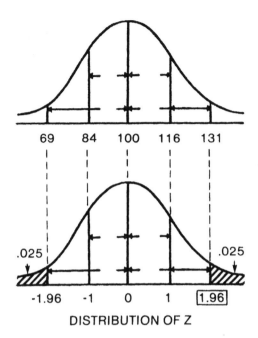

Figure 3.5

ditive law of probability (Section 2.2), the probability of a Z being *either* to the left of −1.96 *or* to the right of +1.96 is .025 + .025 = .05. Transforming back up to *x*, we can say that the probability of someone having an I.Q. outside of 1.96 standard deviations away from the mean (i.e., above 131 or below 69) is .05, or only 5% of the population have values that extreme. (Commonly, the Z value of 1.96 is rounded off to ±2 standard deviations from the mean as corresponding to the cutoff points beyond which lies 5% of the curve, but the accurate value is 1.96.)

A very important use of Z derives from the fact that we can also convert a sample mean (rather than just a single individual value) to a Z score.

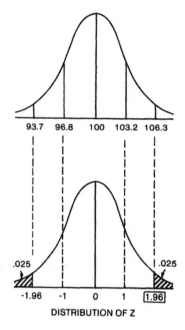

DISTRIBUTION OF MEANS X̄
BASED ON 25 CASES

93.7   96.8   100   103.2   106.3

.025   .025

-1.96   -1   0   1   1.96

DISTRIBUTION OF Z

Figure 3.6

$$Z = \frac{\bar{x} - m}{\sigma_{\bar{x}}}$$

The numerator now is the distance of the sample mean from the population mean and the denominator is the standard deviation of the distribution of means, which is the *standard error of the mean*. This is illustrated in Figure 3.6, where we are considering means based on 25 cases each. The s.e. is $16/\sqrt{25} = 16/5 = 3.2$.

Now we can see that a sample mean of 106.3 corresponds to a Z score of 1.96.

$$Z = \frac{106.3 - 100}{3.2} = 1.96$$

We can now say that the probability that the *mean I.Q. of a group of 25 people* is greater than 106.3 is .025. The probability that such a mean is less than 93.7 is also .025.

A Z score can also be calculated for the *difference between two means*.

$$Z = \frac{(\bar{x}_A - \bar{x}_B) - (m_A - m_B)}{\sigma_{\bar{x}_A - \bar{x}_B}}$$

But $m_A - m_B$ is commonly hypothesized to be 0 so the formula becomes

$$Z = \frac{\bar{x}_A - \bar{x}_B}{\sigma_{\bar{x}_A - \bar{x}_B}}$$

You can see that a *Z score in general is a distance between some value and its mean divided by an appropriate standard error.*

This becomes very useful later on when we talk about confidence intervals in Sections 3.10 to 3.14.

3.9   The t Statistic

Suppose we are interested in sample means and we want to calculate a Z score. We don't know what the population standard deviation is, but if our samples are very large, we can get a good estimate of σ by calculating the standard deviation, s.d., from our sample, and then getting the standard error as usual: s.e. = s.d./$\sqrt{n}$. But often our sample is not large enough. We can still get a standardized score by calculating a value called Student's t:

$$t = \frac{\overline{x} - m}{s.e._{\overline{x}}}$$

It looks just like Z; the only difference is that we calculate it from the sample and it is a small sample.

We can obtain the probability of getting certain t values similarly to the way we obtained probabilities of Z values—from an appropriate table. But it happens, that while the t distribution looks like a normal Z distribution, it is just a little different, thereby giving slightly different probabilities. In fact there are many t distributions (not just one, like for Z). There is a different t distribution for each different sample size. (More will be explained about this in Section 3.10.)

In our example, where we have a mean based on 25 cases, we would need a t value of 2.06 to correspond to a probability of .025 (instead of the 1.96 for the Z distribution). Translating this back to the scale of sample means, if our standard error were 3.2, then the probability would be .025 that we would get a sample mean as large as 106.6 (which is 100 + 2.06 times 3.2), rather than 106.3 (which is 100 + 1.96 times 3.2) as in the Z distribution. This may seem like nit-picking, since the differences are so small. In fact, as the sample size approaches infinity, the t distribution becomes exactly like the Z distribution, but, the differences between Z and t get larger as the sample size gets smaller, and it is always safe to use the t distribution. For example, for a mean based on five cases, the t value would be 2.78 instead of the Z of 1.96. Some t values are tabled in Appendix A. More detailed tables are in standard statistics books.

3.10 Sample Values and Population Values Revisited

All this going back and forth between sample values and population values may be confusing. Here are the points to remember:

(1) We are always interested in estimating population values from samples.
(2) In some of the formulas and terms, we use population values as if we knew what the population values really are. We of

course don't know the actual population values, but if we have very large samples, we can estimate them quite well from our sample data.

(3) For practical purposes, we will generally use and refer to techniques appropriate for small samples, since that is more common and safer (i.e., it doesn't hurt even if we have large samples).

3.11 A Question of Confidence

A confidence interval establishes a range and specifies the probability that this range encompasses the true population mean. For instance, a 95% confidence interval (approximately) is set up by taking the sample mean, $\bar{x}$, plus or minus *two standard errors of the mean*.

95% confidence interval:

$$\bar{x} \pm 2 \; s.e. = \bar{x} \pm 2 \left(\frac{s.d.}{\sqrt{n}} \right)$$

Thus, if we took a random sample of 64 applicants to the Albert Einstein College of Medicine and found their mean I.Q. to be 125, say, (a fictitious figure) we might like to set up a 95% confidence interval to infer what the true mean of the population of applicants really is. The 95% confidence interval is the range between 125–2 s.e. and 125 + 2s.e. We usually phrase this as,

"We are 95% confident that the true mean IQ of Einstein medical school applicants lies within 125 ± 2 s.e."

For the purposes of this example, assume that the standard deviation is 16. (This is not a particularly good assumption since the I.Q. variability of medical school applicants is considerably less than the variability of I.Q. in the population in general.) Under this assumption, we arrive at the following range:

$$125 \pm \frac{2(16)}{\sqrt{64}} = 125 \pm \frac{2(16)}{8} = 125 \pm 4 = 121-129$$

Our statement now is as follows: "The probability is approximately .95 that the true mean I.Q. of Einstein Medical School applicants lies within the range 121–129." (A more rigorous interpretation of this is given in Section 3.11.)

A 99% confidence interval is approximately the sample mean ± 3s.e. In our example this interval would be:

$$125 \pm 3\left[\frac{(16)}{\sqrt{64}}\right] = 125 \pm 6 = 119-131$$

We would then be able to say: "The probability is approximately .99 that the true mean I.Q. of Einstein Medical School applicants lies within the range 119–131."

The "approximately" is because to achieve .95 probability you don't multiply the s.e. by 2 exactly as we did here; we rounded it for convenience. The *exact* multiplying factor depends on how large the sample is. If the sample is very large, greater than 100, we would multiply the s.e. by 1.96 for 95% confidence intervals and by 2.58 for 99% confidence intervals. If the sample is smaller, we should look up the multiplier in tables of t values, which appear in many texts. These t values are different for different "degrees of freedom," explained in Section 3.13, which are related to sample sizes. Some t values are shown in Appendix A. (Also refer back to Section 3.9 for the meaning of t statistics.)

Note that for a given sample size we trade off degree of certainty for size of the interval. We can be more certain that our true mean lies within a wider range but if we want to pin down the range more precisely, we are less certain about it (Figure 3.7). To achieve more precision and maintain a high probability of being correct in estimating the

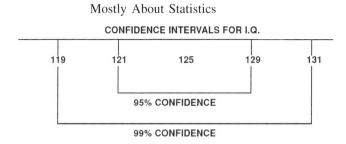

Figure 3.7

range, it is necessary to increase the sample size. The main point here is that when you report a sample mean as an estimate of a population mean, it is most desirable to report the confidence limits.

3.12 Confidence Limits and Confidence Intervals

Confidence limits are the outer boundaries that we calculate and about which we can say: we are 95% confident that these boundaries or limits include the true population mean. The interval between these limits is called the *confidence interval.* If we were to take a large number of samples from the population and calculate the 95% confidence limits for each of them, 95% of the intervals bound by these limits would contain the true population mean. However, 5% would not contain it. Of course, in real life we only take one sample and construct confidence intervals from it. We can never be sure whether the interval calculated from our particular sample is one of the 5% of such intervals that do not contain the population mean. The most we can say is that we are 95% confident it does contain it. As you can see, we never know anything for sure.

If an infinite number of independent random samples were drawn from the population of interest (with replacement), then 95% of the confidence intervals calculated from the samples (with mean $\bar{x}$, and standard error s.e.) will encompass the true population mean m.

Figure 3.8 illustrates the above concepts.

CONFIDENCE INTERVALS

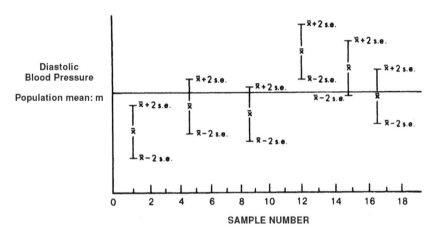

Figure 3.8

3.13 Degrees of Freedom

The t values that we use as the multiplier of the standard error to construct confidence intervals depend on something called the *degrees of freedom* (df), which are related to the sample size. When we have one sample, in order to find the appropriate t value to calculate the confidence limits, we enter the tables with $n - 1$ degrees of freedom, where n is the sample size. An intuitive way to understand the concept of df is to consider that if we calculate the mean of a sample of, say, three values, we would have the "freedom" to vary two of them any way we liked after knowing what the mean is, but the third must be fixed in order to arrive at the given mean. So we only have 2 "degrees of freedom." For example, if we know the mean of three values is 7, we can have the following sets of data:

Value 1:   7   −50

Value 2:   7   +18

Value 3:   <u>7</u>   <u>+53</u>

Sum =   21   21
Mean = $\bar{x} = 7$ $\bar{x} = 7$

In each case, if we know values 1 and 2, then value 3 is determined since the sum of these values must be 21 in order for the mean to be 7. We have "lost" one degree of freedom in calculating the mean.

3.14 Confidence Intervals for Proportions

A proportion can be considered a continuous variable. For example, in the anticoagulant study described in Section 3.1, the proportion of women in the control (placebo-treated) group who survived a heart attack was found to be $89/129 = .69$. A proportion may assume values along the continuum between 0 and 1. We can construct a confidence interval around a proportion in a similar way to constructing confidence intervals around means. The 95% confidence limits for a proportion are $p \pm 1.96$ s.e.$_p$, where s.e.$_p$ is the *standard error of a proportion*.

To calculate the standard error of a proportion, we must first calculate the standard deviation of a proportion and divide it by the square root of n. We define our symbology:

$$s = standard\ deviation\ of\ a\ proportion = \sqrt{pq}$$

$$p = sample\ proportion = \frac{number\ of\ survivors\ in\ control\ group}{total\ number\ of\ women\ in\ control\ group}$$

$$q = 1 - p = \frac{number\ dead\ in\ control\ group}{total\ number\ of\ women\ in\ control\ group}$$

$$s.e._{\cdot p} = \frac{\sqrt{pq}}{\sqrt{n}} = \sqrt{\frac{pq}{n}}$$

In our example of women survivors of a heart attack in the control group, the 95% confidence interval is

$$.69 \pm 1.96\ x\ \sqrt{\frac{(.69) \times (.31)}{129}} = .69 \pm .08$$

And we can make the statement that we are 95% confident that the population proportion of untreated women surviving a heart attack is between .61 and .77 or 61% and 77%. (Remember this refers to the population from which our sample was drawn. We cannot generalize this to all women having a heart attack.)

For 99% confidence limits, we would multiply the standard error of a proportion by 2.58, to get the interval .59 to .80. The multiplier is the Z value that corresponds to .95 for 95% confidence limits or .99 probability for 99% confidence limits.

3.15 Confidence Intervals Around the Difference Between Two Means

We can construct confidence intervals around a difference between means in a similar fashion to which we constructed confidence intervals around a single mean. The 95% confidence limits around the difference between means are given by

$$(\bar{x} - \bar{y}) \pm (t_{df, .95}) (s.e._{\bar{x}-\bar{y}})$$

In words, this is the difference between the two sample means, plus or minus an appropriate t value, times the standard error of the difference; df is the degrees of freedom and .95 says that we look up the t value that pertains to those degrees of freedom and to .95 probability. The degrees of freedom when we are looking at two samples are $n_x + n_y - 2$. This is because we have lost one degree of freedom for each of the two means we have calculated, so our total degrees of freedom is $(n_x - 1) + (n_y - 1) = n_x + n_y - 2$.

As an example consider that we have a sample of 25 female and 25 male medical students. The mean I.Q.s for each sample are

$$\bar{x}_{females} = 130, \quad \bar{x}_{males} = 126, \quad s_{pooled} = 12, \quad df = 48$$

The 95% confidence interval for the mean difference between men and women is calculated as follows:

From t tables, we find that the t value for df = 48 is 2.01

$$\bar{x}_{females} - \bar{x}_{males} \pm 2.01 \times \sqrt{s_p \left(\frac{1}{n_x} + \frac{1}{n_y} \right)} =$$

$$(130 - 126) \pm 2.01 \times \sqrt{12(1/25 + 1/25)} = 4 \pm 6.8$$

The interval then is -2.8 to 10.8, and we are 95% certain it includes the true mean difference between men and women. This interval includes 0 difference, so we would have to conclude that the difference in I.Q. between men and women may be zero.

3.16 Comparisons Between Two Groups

A most common problem that arises is the need to compare two groups on some dimension. We may wish to determine, for instance, whether (1) administering a certain drug lowers blood pressure, or (2) drug A is more effective than drug B in lowering blood sugar levels, or (3) teaching first-grade children to read by method I produces higher reading achievement scores at the end of the year than teaching them to read by method II.

3.17 Z-Test for Comparing Two Proportions

As an example we reproduce here the table in Section 3.1 showing data from a study on anticoagulant therapy.

Observed Frequencies

	Control	Treated	
Lived	89	223	312
Died	40	39	79
Total	129	262	391

If we wish to test whether the proportion of women surviving a heart attack in the treated group differs from the proportion surviving in the control group we set up our null hypothesis as

H_0:　　$P_1 = P_2$ or $P_1 - P_2 = 0$;　　P_1 = proportion surviving in treated population

P_2 = proportion surviving in control population

H_A:　　$P_1 - P_2 \neq 0$　　(the difference does not equal 0)

We calculate

$$Z = \frac{p_1 - p_2}{s.e._{p_1 - p_2}}$$

$$p_1 = \frac{223}{262} = .85, \qquad q_1 = 1 - p_1 = .15, \qquad n_1 = 262$$

$$p_2 = \frac{89}{129} = .69, \qquad q_2 = 1 - p_2 = .31, \qquad n_2 = 129$$

Thus, the numerator of $Z = .85 - .69 = .16$.

The denominator =
standard error of the difference between two proportions =

$$s.e._{(p_1 - p_2)} = \sqrt{\hat{p}\hat{q}\left(\frac{1}{n_1} + \frac{1}{n_2}\right)}$$

where $\hat{p}$ and $\hat{q}$ are pooled estimates based on both treated and control group data. We calculate it as follows:

$$\hat{p} = \frac{n_1 p_1 + n_2 p_2}{n_1 + n_2} = \frac{number\ of\ survivors\ in\ treated + control}{total\ number\ of\ patients\ in\ treated + control}$$

$$= \frac{262(.85) + 129(.69)}{262 + 129} = \frac{223 + 89}{391} = .80$$

$$\hat{q} = 1 - \hat{p} = 1 - .80 = .20$$

$$s.e._{(p_1-p_2)} = \sqrt{(.80)(.20)\left(\frac{1}{262} + \frac{1}{129}\right)} = .043$$

$$Z = \frac{.85 - .69}{.043} = 3.72$$

We must now look to see if this value of Z exceeds the *critical value*. *The critical value is the minimum value of the test statistics that we must get in order to reject the null hypothesis at a given level of significance.*

The *critical value of Z* that we need to reject H_O at the .05 level of significance is 1.96. The value we obtained is 3.74. This is clearly a large enough Z to reject H_O at the .01 level at least. The critical value for Z to reject H_O at the .01 level is 2.58.

Note that we came to the same conclusion using the chi-square test in Section 3.1. In fact $Z^2 = \chi^2 = (3.74)^2 = 13.99$ and the uncorrected chi-square we calculated was 13.94 (the difference is due to rounding errors). Of course the critical values of χ^2 and Z have to be looked up in their appropriate tables. Some values appear in Appendix A.

3.18 t-Test for the Difference Between Means of Two Independent Groups: Principles

When we wanted to compare two groups on some measure that was a discrete or categorical variable, like mortality in two groups, we used the chi-square test, described in Section 3.1. Or we could use a test between proportions as described in Section 3.17. We now discuss a method of comparing two groups when the measure of interest is a continuous variable.

Let us take as an example the comparison of the ages at first pregnancy of two groups of women: those who are lawyers and those who are paralegals. Such a study might be of sociological interest, or it might be of interest to law firms, or perhaps to a baby foods company that is seeking to focus its advertising strategy more effectively.

Assuming we have taken proper samples of each group, we now have two sets of values: the ages of the lawyers (group A) and the ages of the paralegals (group B), and we have a mean age for each sample. We set up our null hypothesis as follows:

H_0:   "The mean age of the population of lawyers from which we have drawn sample A is the same as the mean age of the population of paralegals from which we have drawn sample B."

Our alternate hypothesis is

H_A: "The mean ages of the two populations we have sampled are different."

In essence then, *we have drawn samples on the basis of which we will make inferences about the populations from which they came.* We are subject to the same kinds of type I and type II errors we discussed before.

The general approach is as follows. We know there is variability of the scores in group A around the mean for group A and within group B around the mean for group B, simply because even within a given population, people vary. What we want to find is whether the variability between the two sample means around the grand mean of all the scores is greater than the variability of the ages within the groups around their own means. If there is as much variability within the groups as between the groups, then they probably come from the same population.

The appropriate test here is the t-test. We calculate a value known as t, which is equal to the difference between the two sample means divided by an appropriate standard error. The appropriate standard error is called the standard error of the difference between two means and is written as

$$s.e._{\bar{x}_1 - \bar{x}_2}$$

The distribution of t has been tabulated and from the tables we can obtain the probability of getting a value of t as large as the one we actually obtained under the assumption that our null hypothesis (of no difference between means) is true. If this probability is small (i.e., if it is unlikely that by chance alone we would get a value of t that large if the null hypothesis were true) we would reject the null hypothesis and accept the alternate hypothesis that there really is a difference between the means of the populations from which we have drawn the two samples.

3.19 How to Do a t-Test: An Example

Although t-tests can be easily performed on personal computers, an example of the calculations and interpretation is given below. This statistical test is performed to compare the means of two groups under the assumption that both samples are random, independent, and come from normally distributed populations with unknown but equal variances.

Null Hypothesis: $m_A = m_B$, or the equivalent: $m_A - m_B = 0$.

Alternate Hypothesis: $m_A \neq m_B$, or the equivalent: $m_A - m_B \neq 0$.

[Note: When the alternate hypothesis does not specify the direction of the difference (by stating for instance that m_A is greater than m_B) but simply says the difference *does not equal 0*, it is called a two-tailed test. When the direction of the difference is specified, it is called a one-tailed test. More on this topic appears in Section 5.4.]

$$t = \frac{(\bar{x}_A - \bar{x}_B)}{s_{\bar{x}_A - \bar{x}_B}}$$

	Ages of Sample A			Ages of Sample B	
x_i	$(x_i - \bar{x}_A)$	$(x_i - \bar{x}_A)^2$	x_i	$(x_i - \bar{x}_B)$	$(x_i - \bar{x}_B)^2$
28	−3	9	24	2.4	5.76
30	−1	1	25	3.4	11.56
27	−4	16	20	−1.6	2.56
32	1	1	18	−3.6	12.96
34	3	9	21	−0.6	0.36
36	5	25	$\Sigma = 108$	$\Sigma = 0$	$\Sigma = 33.20$
30	−1	1			
$\Sigma = 217$	$\Sigma = 0$	$\Sigma = 62$			

$$Mean_A = \bar{x}_A = \frac{\Sigma x_i}{n} = \frac{217}{7} = 31; \quad Mean_B = \bar{x}_B = \frac{\Sigma x_i}{n} = \frac{108}{5} = 21.6$$

(The subscript i refers to the ith score and is a convention used to indicate that we sum over all the scores.)

The numerator of t is the difference between the two means:

$$31 - 21.6 = 9.4$$

To get the denominator of t we need to calculate the standard error of the difference between means, which we do as follows:

First we get the pooled estimate of the standard deviation. We calculate:

$$s_p = \sqrt{\frac{\Sigma(x_i - \bar{x}_A)^2 + \Sigma(x_i - \bar{x}_B)^2}{n_A + n_B - 2}} = \sqrt{\frac{62 + 33.20}{7 + 5 - 2}}$$

$$= \sqrt{\frac{95.20}{10}} = \sqrt{9.52} = 3.09$$

$$S_{\bar{x}_A - \bar{x}_B} = S_p \sqrt{\frac{1}{n_A} + \frac{1}{n_B}} = 3.09 \sqrt{\frac{1}{7} + \frac{1}{5}} = 3.09 \sqrt{.3428} = 3.09 \times .5854 = 1.81$$

$$t = \frac{\bar{x}_A - \bar{x}_B}{S_{\bar{x}_A - \bar{x}_B}} = \frac{9.4}{1.81} = 5.19$$

This t is significant at the .001 level, which means that you would get a value of t as high as this one or higher only 1 time out of a thousand by chance if the null hypothesis were true. So we reject the null hypothesis of no difference, accept the alternate hypothesis, and conclude that the lawyers are older at first pregnancy than the paralegals.

3.20 Matched Pair t-Test

If you have a situation where the scores in one group correlate with the scores in the other group, you cannot use the regular t-test since that assumes the two groups are independent. This situation arises when you take two measures on the same individual. For instance, suppose group A represents reading scores of a group of children taken at time 1. These children have then been given special instruction in reading over a period of six months and their reading achievement is again measured to see if they accomplished any gains at time 2. In such a situation you would use a matched pair t-test.

Child	A Initial reading scores of children	B Scores of same children after 6 months' training	$d = B - A$	$d - \bar{d}$	$(d - \bar{d})^2$
(1)	60	62	2	1.4	1.96
(2)	50	54	4	3.4	11.56
(3)	70	70	0	−0.6	0.36
(4)	80	78	−2	−2.6	6.76
(5)	75	74	−1	−1.6	2.56
		Sum	3	0	23.20
	Mean difference = $\bar{d}$ = 3/5 = 0.60				

Null Hypothesis: Mean difference = 0.

Alternate Hypothesis: Mean difference is greater than 0.

$$t = \frac{\overline{d}}{s_{\overline{d}}}; \quad s_{\overline{d}} = \frac{s}{\sqrt{n}}$$

$$s = \sqrt{\frac{\Sigma(d - \overline{d})^2}{n-1}} = \sqrt{\frac{23.20}{4}} = \sqrt{5.8} = 2.41$$

$$s_{\overline{d}} = \frac{2.41}{\sqrt{5}} = \frac{2.41}{2.23} = 1.08$$

$$t = \frac{.60}{1.08} = .56$$

This t is not significant, which means that we do not reject the null hypothesis and conclude that the mean difference in reading scores could be zero; that is, the six months' reading program may not be effective. (Or it may be that the study was not large enough to detect a difference, and we have committed a type II error.)

When the actual difference between matched pairs is not in itself a meaningful number, but the researcher can *rank* the difference scores (as being larger or smaller for given pairs). The appropriate test is the Wilcoxon matched-pairs rank sums test. This is known as a *nonparametric test*, and along with other such tests is described with exquisite clarity in the classic book by Sidney Siegel, *Nonparametric Statistics for the Behavioral Sciences* (listed in the Suggested Readings).

3.21 When Not to Do a Lot of t-Tests: The Problem of Multiple Tests of Significance

A t-test is used for comparing the means of two groups. When there are three or more group means to be compared, the t-test is not appropriate. To understand why, we need to invoke our knowledge of combining probabilities from Section 2.2.

Suppose you are testing the effects of three different treatments for high blood pressure. Patients in one group A receive one medication, a diuretic; patients in group B receive another medication, a beta-blocker; and patients in group C receive a placebo pill. You want to know whether either drug is better than placebo in lowering blood pressure and if the two drugs are different from each other in their blood pressure lowering effect.

There are three comparisons that can be made: group A versus group C (to see if the diuretic is better than placebo), group B versus group C (to see if the beta-blocker is better than the placebo), and group A versus group B (to see which of the two active drugs has more effect). We set our significance level at .05, that is, we are willing to be *wrong* in rejecting the null hypothesis of no difference between two means, with a probability of .05 or less (i.e., our probability of making a type I error must be no greater than .05). Consider the following:

Comparison	Probability of type I error	Probability of *not* making a type I error = $1 - P$ (type I error)
1. A vs. C	.05	$1 - .05 = .95$
2. B vs. C	.05	$1 - .05 = .95$
3. A vs. B	.05	$1 - .05 = .95$

The probability of *not* making a type I error in the first comparison *and* not making it in the second comparison *and* not making it in the third comparison = .95 x .95 x .95 = .86. We are looking here at the *joint* occurrence of three events (the three ways of *not* committing a type I error) and we combine these probabilities by multiplying the individual probabilities. (Remember, when we see "and" in the context of combining probabilities, we multiply, when we see "or" we add.) So now, we know that the overall probability of *not* committing a type I error in any of the three possible comparisons is .86. Therefore, the probability of committing such an error is 1—the probability of not committing it, or $1 - .86 = .14$. Thus, the overall probability of a type I error would be considerably greater than the .05 we specified as desirable. In actual fact, the numbers are a little different because the three comparisons are not independent events, since the same groups are used in more than one comparison, so combining probabilities in this situation would not involve the simple multiplication rule for the joint occurrence of independent events. However, it is close enough to illustrate the point that making multiple comparisons in the same experiment results in quite a different significance level (.14 in this example) than the one we chose (.05). When there are more than three groups to compare, the situation gets worse.

3.22 Analysis of Variance: Comparison Among Several Groups

The appropriate technique for analyzing continuous variables when there are three or more groups to be compared is the analysis of variance, commonly referred to as ANOVA. An example might be comparing the blood pressure reduction effects of the three drugs.

3.23 Principles

The principles involved in the analysis of variance are the same as those in the t-test. Under the null hypothesis we would have the following situation: there would be one big population and if we picked samples of a given size from that population we would have a bunch of sample means that would vary due to chance around the grand mean of the whole population. If it turns out they vary around the grand mean more than we would expect just by chance alone, then perhaps something other than chance is operating. Perhaps they don't all come from the same population. Perhaps something distinguishes the groups we have picked. We would then reject the null hypothesis that all the means are equal and conclude the means are different from each other by more than just chance. Essentially, we want to know if the variability of all the groups means is substantially greater than the variability within each of the groups around their own mean.

We calculate a quantity known as the *between-groups variance*, which is the variability of the group means around the grand mean of all the data. We calculate another quantity called the *within-groups variance*, which is the variability of the scores within each group around its own mean. One of the assumptions of the analysis of variance is that the extent of the variability of individuals within groups is the same for each of the groups, so we can pool the estimates of the individual within group variances to obtain a more reliable estimate of overall within-groups variance. If there is as much variability of individuals *within* the groups as there is variability of means *between* the groups, the means probably come from the same population, which would be consistent with the hypothesis of no true difference among means, that is, we could not reject the null hypothesis of no difference among means.

The ratio of the between-groups variance to the within-groups variance is known as the F ratio. Values of the F distribution appear in tables in many statistical texts and if the obtained value from our experiment is greater than the *critical value* that is tabled, we can then reject the hypothesis of no difference.

There are different critical values of F depending on how many groups are compared and on how many scores there are in each group. To read the tables of F, one must know the two values of degrees of freedom (df). The df corresponding to the between-groups variance, which is the numerator of the F ratio, is equal to $k - 1$, where k is the number of groups. The df corresponding to the denominator of the F ratio, which is the within-groups variance, is equal to $k \times (n - 1)$, that is, the number of groups times the number of scores in each group minus one. For example, if in our hypertension experiment there are 100 patients in each of the three drug groups, then the numerator degrees of freedom would be $3 - 1 = 2$, and the denominator degrees of freedom would be $3 \times 99 = 297$. An F ratio would have to be at least 3.07 for a significance level of .05. If there were four groups being compared then the numerator degrees of freedom would be 3, and the critical value of F would need to be 2.68. If there is not an equal number of individuals in each group, then the denominator degrees of freedom is $(n_1 - 1) + (n_2 - 1) + (n_3 - 1)$.

We will not present here the actual calculations necessary to do an F test because nowadays these are rarely done by hand. There are a large number of programs available for personal computers that can perform F tests, t-tests, and most other statistical analyses. However, shown below is the kind of output that can be expected from these programs. Shown are summary data from the TAIM study (Trial of Antihypertensive Interventions and Management). The TAIM study was designed to evaluate the effect of diet and drugs, used alone or in combination with each other, to treat overweight persons with mild hypertension (high blood pressure).[10,11]

The next table shows the mean drop in blood pressure after six months of treatment with each drug, the number of people in each group, and the standard deviation of the *change* in blood pressure in each group.

Drug group	n	Mean drop (in diastolic blood pressure units after 6 months of treatment)	Standard deviation
A. Diuretic	261	12.1	7.9
B. Beta-blocker	264	13.5	8.2
C. Placebo	257	9.8	8.3

The next table results from an analysis of variance of the data from this study. It is to be interpreted as follows:

ANOVA					
Source of variation	Degrees of freedom	Sum of squares	Mean square	F ratio	$P_2 > F$
Between groups	2	1776.5	888.2	13.42	.0001
Within groups	779	5256.9	66.2		
	781				

The mean square is the sum of squares divided by the degrees of freedom. For between-groups, it is the variation of the group means around the grand mean, while for within-groups it is the pooled estimate of the variation of the individual scores around their respective group means. The within-groups mean square is also called the error mean square. (An important point is that the square root of the error mean square is the pooled estimate of the within-groups standard deviation. In this case it is $\sqrt{66.2} = 8.14$. It is roughly equivalent to the average standard deviation.) F is the ratio of the between to the within mean squares; in this example it is $888.2/66.2 = 13.42$.

The F ratio is significant at the .0001 level, so we can reject the null hypothesis that *all* group means are equal. However, we do not know where the difference lies. Is group A different from group C but not from group B? We should not simply make all the pairwise comparisons possible because of the problem of multiple comparisons discussed above. But there are ways to handle this problem. One of them is the Bonferroni procedure, described in the next section.

3.24 Bonferroni Procedure: An Approach to Making Multiple Comparisons

This is one way to handle the problem of multiple comparisons. The Bonferroni procedure implies that if for example we make five comparisons, the probability that *none* of the five p values falls below .05 is at least $1 - (5 \times .05) = .75$ when the null hypothesis of equal means is really true. That means that there is a probability of up to .25 that at least one p value will reach the .05 significance level by chance alone *even if the treatments really do not differ*. To get around this, we divide the chosen overall significance level by the number of two-way comparisons to be made, consider this value to be the significance level for any single comparison, and reject the null hypothesis of no difference only if it achieves this new significance level.

For example, if we want an overall significance level of .05 and we will make three comparisons between means, we would have to achieve $.05/3 = .017$ level in order to reject the null hypothesis and conclude there is a difference between the two means. A good deal of self-discipline is required to stick to this procedure and not declare a difference between two means as unlikely to be due to chance if the particular comparison has significance at $p = .03$, say, instead of .017. The Bonferroni procedure does not require a prior F test. Let us apply the Bonferroni procedure to our data.

First we compare each of the drugs to placebo. We calculate the t for the difference between means of group A versus group C.

$$t = \frac{\bar{x}_A - \bar{x}_C}{s.e._{\bar{x}_A - \bar{x}_C}}$$

$$s.e._{\bar{x}_A - \bar{x}_C} = s_p \sqrt{\frac{1}{n_A} + \frac{1}{n_C}}$$

$$\frac{12.1 - 9.8}{8.14\sqrt{\frac{1}{261} + \frac{1}{257}}} = \frac{2.3}{.715} = 3.22$$

$$p = .0014$$

Note that we use 8.14 as s pooled. We obtained this from the analysis of variance as an estimate of the common standard deviation. The degrees of freedom to enter the t tables are $261 + 257 - 2 = 516$.

It turns out that the probability of getting such a high t value by chance is only .0014. We can safely say the diuretic reduces blood pressure more than the placebo. The same holds true for the comparison between the beta-blocker and placebo. Now let us compare the two drugs, B versus A:

$$t = \frac{13.5 - 12.1}{8.14\sqrt{\dfrac{1}{264} + \dfrac{1}{261}}} = \frac{1.4}{.711} = 1.97$$

The p value corresponding to this t value is .049. It might be tempting to declare a significant difference at the .05 level, but remember the Bonferroni procedure requires that we get a p value of .017 or less for significance adjusted for multiple comparisons. The critical value of t corresponding to $p = .017$ is 2.39 and we only got a t of 1.97. Recently,[12] there has been some questioning of the routine adjustment for multiple comparisons on the grounds that we thereby may commit more type II errors and miss important effects. In any case p levels should be reported so that the informed reader may evaluate the evidence.

3.25 Analysis of Variance When There Are Two Independent Variables: The Two-Factor ANOVA

The example above is referred to as the one-way ANOVA because you can divide all the scores in one way only, by the drug group to which patients were assigned. The drug group is called a "factor" and this factor has three levels, meaning there are three categories of drug. There may, however, be another factor that classifies individuals, and in that case we would have a two-way, or a two-factor, ANOVA. In the experiment we used as an example, patients were assigned to one of the three drugs noted above, as well as to one of three diet regimens—weight reduction, sodium (salt) restriction, or no change from their usual diet, which is analogous to a placebo diet condition. The diagram below illustrates this two-factor design, and the mean drop in blood pressure in each group, as well as the numbers of cases in each group, which are shown in parenthesis.

Drug	Diet			Total
	Usual	Weight reduction	Sodium restriction	
Diuretic	10.2 (87)	14.5 (86)	11.6 (88)	12.1 (261)
Beta-blocker	12.8 (86)	15.2 (88)	12.6 (90)	13.5 (264)
Placebo	8.7 (89)	10.8 (89)	10.1 (79)	9.8 (257)
Total	10.5 (262)	13.5 (263)	11.5 (257)	

Now we are interested in comparing the three means that represent change in blood pressure in the drug groups, the three means that represent changes in the diet groups, and the interaction between drug and diet. We now explain the concept of interaction.

3.26 Interaction Between Two Independent Variables

Interaction between two independent variables refers to differences in the effect of one variable depending on the level of the second variable. For example, maybe one drug produces better effects when combined with a weight-reduction diet than when combined with a sodium-restricted diet. There may not be a significant effect of that drug when all diet groups are lumped together but if we look at the effects separately for each diet group we may discover an interaction between the two factors: diet and drug.

The diagrams below illustrate the concept of interaction effects. WR means weight reduction and SR means sodium (salt) restriction.

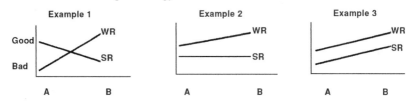

In example 1 drug B is better than drug A in those under weight reduction but in those under salt restriction drug A is better than drug B. If we just compared the average for drug A, combining diets, with the average for drug B, we would have to say there is no difference between drug A and drug B, but if we look at the two diets separately we see quite different effects of the two drugs.

In example 2, there is no difference in the two drugs for those who restrict salt, but there is less effect of drug A than drug B for those in weight reduction.

In example 3, there is no interaction; there is an equal effect for both diets: the two lines are parallel; their slopes are the same. Drug B is better than drug A both for those in weight reduction and salt restriction.

3.27 Example of a Two-Way ANOVA

Next is a table of data from the TAIM study showing the results of a *two-way analysis of variance:*

Two-Way ANOVA From TAIM Data

Source	DF	ANOVA sum of squares	Mean square	F value	Probability
Drug group	2	1776.49	888.25	13.68	.0001
Diet group	2	1165.93	582.96	8.98	.0001
Drug × diet	4	214.50	53.63	0.83	.509
Error	773	50,185.46	64.93		

Note that the error mean square here is 64.93 instead of 66.9 when we did the one-way analysis. That is because we have explained some of the error variance as being due to diet effects and interaction effects (we have "taken out" these effects from the error variance). Thus, 64.93 represents the variance due to pure error, or the unexplained variance. Now we can use the square root of this which is 8.06 as the estimate of the common standard deviation. We explain the results as follows: There is a significant effect of drug ($p = .0001$) and a significant effect of diet ($p = .0001$), but no interaction of drug by diet ($p = .509$).

We have already made the three pairwise comparisons, by t-tests for the difference between two means among drugs (i.e., placebo vs. diuretic, placebo vs. beta-blocker, and diuretic vs. beta-blocker). We can do the same for the three diets. Their mean values are displayed below:

Diet group	n	Mean drop in diastolic blood pressure	Standard deviation
Weight reduction	263	13.5	8.3
Sodium restriction	257	11.5	8.3
Usual diet	262	10.5	8.0
(Pooled estimate of s.d. = 8.06)			

If we did t-tests, we would find that weight reduction is better than usual diet ($p = .0000$), but sodium restriction shows no significant improvement over usual diet ($p = .16$).

Weight reduction when compared with sodium restriction is also significantly better with $p = .005$, which is well below the $p = .017$ required by the Bonferroni procedure. (The t for this pairwise comparison is 2.83, which is above the critical value of 2.39.)

3.28 Kruskal–Wallis Test to Compare Several Groups

The analysis of variance is valid when the variable of interest is continuous, comes from a normal distribution, that is, the familiar bell-shaped curve, and the variances within each of the groups being compared are essentially equal. Often, however, we must deal with situations when we want to compare several groups on a variable that does not meet all of the above conditions. This might be a case where we can say one person is better than another, but we can't say exactly how much better. In such a case we would rank people and compare the groups by using the Kruskal–Wallis test to determine if it is likely that all the groups come from a common population. This test is analogous to the one-way analysis of variance but instead of using the original scores, it uses the *rankings of* the scores. It is called a *non-parametric test*. This test is available in many computer programs, but an example appears in Appendix C.

3.29 Association and Causation: The Correlation Coefficient

A common class of problems in the accumulation and evaluation of scientific evidence is the assessment of association of two variables. Is there an association between poverty and drug addiction? Is emotional stress associated with cardiovascular disease?

To determine association, we must first quantify both variables. For instance, emotional stress may be quantified by using an appropriate psychological test of stress or by clearly defining, evaluating, and rating on a scale the stress factor in an individual's life situation, whereas hypertension (defined as a blood pressure reading) may be considered as the particular aspect of cardiovascular disease to be studied. When variables have been quantified, a measure of association needs to be calculated to determine the strength of the relationship. One of the most common measures of association is the *correlation coefficient, r*, which is a number derived from the data that can vary between −1 and +1. (The method of calculation appears in Appendix D.) When $r = 0$ it means there is no association between the two variables. An example of this might be the correlation between blood pressure and the number of hairs on the head. When $r = +1$, a perfect positive correlation, it means there is a direct relationship between the two variables: an individual who has a high score on one variable also has a high score on the other, and the score on one variable can be exactly predicted from the score on the other variable. This kind of correlation exists only in deterministic models, where there is really a functional relationship. An example might be the correlation between age of a tree and the number of rings it has. A correlation coefficient of -1 indicates a perfect inverse relationship, where a high score on one variable means a low score on the other and where, as in perfect positive correlation, there is no error of measurement. Correlation coefficients between 0 and +1 and between 0 and -1 indicate varying strengths of associations.

These correlation coefficients apply when the basic relationship between the two variables is linear. Consider a group of people for each of whom we have a measurement of weight against height; we will find that we can draw a straight line through the points. There is a linear association between weight and height and the correlation coefficient would be positive but less than 1.

The diagrams in Figure 3.9 illustrate representations of various correlation coefficients.

3.30 How High Is High?

The answer to this question depends upon the field of application as well as on many other factors. Among psychological variables, which are difficult to measure precisely and are affected by many other variables, the correlations are generally (though not necessarily) lower than among biological variables where more accurate measurement is possible. The following example may give you a feel for the orders of magnitude. The correlations between verbal aptitude and nonverbal aptitude, as measured for Philadelphia schoolchildren by standardized national tests, range from .44 to .71 depending on race and social class of the groups.[13]

3.31 Causal Pathways

If we do get a significant correlation, we then ask what situations could be responsible for it? Figure 3.10 illustrates some possible strucstruc-

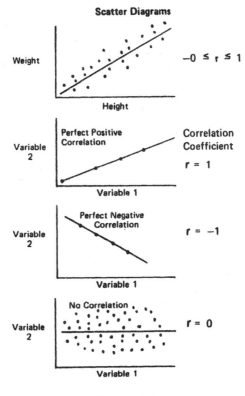

Figure 3.9

tural relationships that may underlie a significant correlation coefficient, as suggested by Sokal and Rohlf. [14] We consider two variables, W
(weight gain) and B (blood pressure) , and let r_{WB} represent the correlation between them. Note that only in diagrams (1), (2), and (6) does the
correlation between W and B arise due to a causal relationship between
the two variables. In diagram (1), W entirely determines B; in diagram
(2), W is a partial cause of B; in diagram (6), W is one of several determinants of B. In all of the other structural relationships, the correlation between W and B arises due to common influences on both variables. Thus, it must be stressed that *the existence of a correlation
between two variables does not necessarily imply causation.* Correla-

tions may arise because one variable is the partial cause of another or the two correlated variables have a common cause. Other factors, such as sampling, the variation in the two populations, and so on, affect the size of the correlation coefficient also. Thus, care must be taken in interpreting these coefficients.

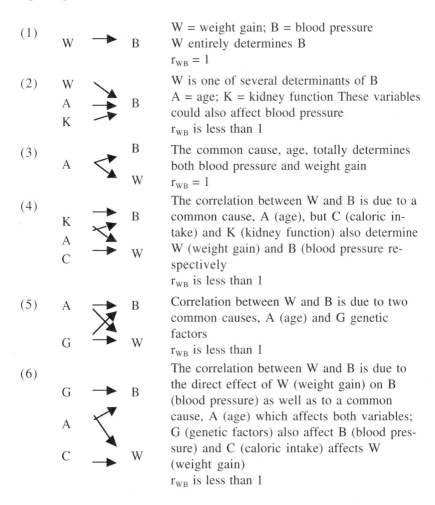

(1) W → B

W = weight gain; B = blood pressure
W entirely determines B
$r_{WB} = 1$

(2) W, A, K → B

W is one of several determinants of B
A = age; K = kidney function These variables could also affect blood pressure
r_{WB} is less than 1

(3) A → B, W

The common cause, age, totally determines both blood pressure and weight gain
$r_{WB} = 1$

(4) K, A, C → B, W

The correlation between W and B is due to a common cause, A (age), but C (caloric intake) and K (kidney function) also determine W (weight gain) and B (blood pressure respectively
r_{WB} is less than 1

(5) A, G → B, W

Correlation between W and B is due to two common causes, A (age) and G genetic factors
r_{WB} is less than 1

(6) G, A, C → B, W

The correlation between W and B is due to the direct effect of W (weight gain) on B (blood pressure) as well as to a common cause, A (age) which affects both variables; G (genetic factors) also affect B (blood pressure) and C (caloric intake) affects W (weight gain)
r_{WB} is less than 1

Figure 3.10

3.32 Regression

Note that in Figure 3.9 we have drawn lines that seem to best fit the data points. These are called *regression lines*. They have the following form:

$Y = a + bX$. In the top scattergram labeled (a), Y is the dependent variable weight and X, or height, is the independent variable. We say that weight is a function of height. The quantity a is the intercept. It is where the line crosses the Y axis. The quantity b is the slope and it is the rate of change in Y for a unit change in X. If the slope is 0, it means we have a straight line parallel to the x axis, as in the illustration (d). It also means that we cannot predict Y from a knowledge of X since there is no relationship between Y and X. If we have the situation shown in scattergrams (b) or (c), we know exactly how Y changes when X changes and we can perfectly predict Y from a knowledge of X with no error. In the scattergram (a), we can see that as X increases Y increases but we can't predict Y perfectly because the points are scattered around the line we have drawn. We can, however, put confidence limits around our prediction, but first we must determine the form of the line we should draw through the points. We must estimate the values for the intercept and slope. This is done by finding the "best-fit line."

The line that fits the points best has the following characteristics: if we take each of the data points and calculate its vertical distance from the line and then square that distance, the sum of those squared distances will be smaller than the sum of such squared distances from any other line we might draw. This is called the *least-squares* fit. Consider the data below where Y could be a score on one test and X could be a score on another test.

	Score	
Individual	X	Y
A	5	7
B	8	4
C	15	8
D	20	10
E	25	14

The calculations to determine the best-fit line are shown in Appendix D. However, most statistical computer packages for personal computers provide a linear regression program that does these calculations. Figure 3.11 illustrates these points plotted in a scattergram and shows the least-squares line.

The equation for the line is $Y = 2.76 + .40\ X$. The intercept a is 2.76 so that the line crosses the y axis at $Y = 2.76$. The slope is .40. For example, we can calculate a predicted Y for $X = 10$ to get

$$Y = 2.76 + (.40)(10) = 2.76 + 4 = 6.76$$

The d_i's are distances from the points to the line. It is the sum of these squared distances that is smaller for this line than it would be for any other line we might draw.

The correlation coefficient for these data is .89. The square of the correlation coefficient, r^2, can be interpreted as the proportion of the variance in Y that is explained by X. In our example, $.89^2 = .79$; thus

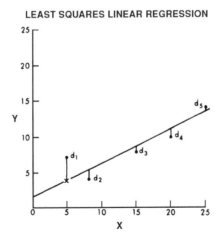

Figure 3.11

79% of the variation of Y is explainable by the variable X, and 21% is unaccounted for.

3.33 The Connection Between Linear Regression and the Correlation Coefficient

The correlation coefficient and the slope of the linear regression line are related by the formulas

$$r = b\frac{s_x}{s_y}, \qquad\qquad b = r\frac{s_y}{s_x}$$

where s_x is the standard deviation of the X variable, s_y is the standard deviation of the Y variable, b is the slope of the line, and r is the correlation coefficient.

3.34 Multiple Linear Regression

When we have two or more independent variables and a continuous dependent variable, we can use multiple regression analysis. The form this takes is

$$Y = a + b_1 X_1 + b_2 X_2 + b_3 X_3 + \ldots + b_k X_k$$

For example, Y may be blood pressure and X_1 may be age, X_2 may be weight, X_3 may be family history of high blood pressure. We can have as many variables as appropriate, where the last variable is the kth variable. The b_i's are *regression coefficients*. Note that family history of high blood pressure is not a continuous variable. It can either be yes or no. We call this a dichotomous variable and we can use it as any other variable in a regression equation by assigning a number to each of the two possible answers, usually by making a yes answer = 1 and a no answer = 0. Statistical computer programs usually include multiple linear regression.

An example from the TAIM study follows and is meant only to give you an idea of how to interpret a multiple regression equation. This analysis pertains to the group of 89 people who were assigned to a placebo drug and a weight-reduction regimen. The dependent variable is change in blood pressure.

The independent variables are shown below:

Variable	Coefficient: b_i	p
Intercept	−15.49	.0016
Age	.077	.359
Race 1 = Black	4.22	.021
0 = Nonblack		
Sex 1 = Male	1.50	.390
0 = Female		
Pounds lost	.13	.003

Note: Sex is coded as 1 if male and 0 if female; race is coded as 1 if black and 0 if nonblack; p is used to test if the coefficient is significantly different from 0. The equation, then, is

$$\text{change in blood pressure} =$$

$$-15.49 + .077(\text{age}) + 4.22(\text{race}) + 1.5(\text{sex}) + .13(\text{change in weight})$$

Age is not significant ($p = .359$), nor is sex ($p = .390$). However, race is significant ($p = .021$), indicating that blacks were more likely than nonblacks to have a drop in blood pressure while simultaneously controlling for all the other variables in the equation. Pounds lost is also significant, indicating that the greater the weight loss the greater was the drop in blood pressure.

3.35 Summary So Far

Investigation of a scientific issue often requires statistical analysis, especially where there is variability with respect to the characteristics of interest. The variability may arise from two sources: the characteristic may be inherently variable in the population and/or there may be error of measurement.

In this chapter we have pointed out that in order to evaluate a program or a drug, to compare two groups on some characteristic, to conduct a scientific investigation of any issue, it is necessary to quantify the variables.

Variables may be quantified as discrete or as continuous and there are appropriate statistical techniques that deal with each of these. We have considered here the chi-square test, confidence intervals, Z-test, t-test, analysis of variance, correlation, and regression. We have pointed out that in hypothesis testing we are subject to two kinds of errors: the error of rejecting a hypothesis when in fact it is true, and the error of accepting a hypothesis when in fact it is false. The aim of a well-designed study is to minimize the probability of making these types of errors. Statistics will not substitute for good experimental design, but it is a necessary tool to evaluate scientific evidence obtained from well-designed studies.

Philosophically speaking, statistics is a reflection of life in two important respects: (1) as in life, we can never be certain of anything (but in statistics we can put a probability figure describing the degree of our uncertainty), and (2) all is a trade-off—in statistics, between certainty and precision, or between two kinds of error; in life, well, fill in your own trade-offs.

Chapter 4
MOSTLY ABOUT EPIDEMIOLOGY

Medicine to produce health has to examine disease; and music to create harmony, must investigate discord.

Plutarch
A.D. 46–120

4.1 The Uses of Epidemiology

Epidemiology may be defined as the study of the distribution of health and disease in *groups of people* and the study of the factors that influence this distribution. Modern epidemiology also encompasses the evaluation of diagnostic and therapeutic modalities and the delivery of health care services. There is a progression in the scientific process (along the dimension of increasing credibility of evidence), from casual observation, to hypothesis formation, to controlled observation, to experimental studies. Figure 4.1 is a schematic representation of the uses of epidemiology. The tools used in this endeavor are in the province of epidemiology and biostatistics. The techniques used in these disciplines enable "medical detectives" both to uncover a medical problem, to evaluate the evidence about its causality or etiology, and to evaluate therapeutic interventions to combat the problem.

Descriptive epidemiology provides information on the pattern of diseases, on "who has what and how much of it," information that is essential for health care planning and rational allocation of resources. Such information may often uncover patterns of occurrence suggesting etiologic relationships and can lead to preventive strategies. Such studies are usually of the cross-sectional type and lead to the formation of hypotheses that can then be tested in case-control, prospective, and experimental studies. Clinical trials and other types of controlled studies serve to evaluate therapeutic modalities and other means of interventions and thus ultimately determine standards of medical practice, which in turn have impact on health care planning decisions. In the following section, we will consider selected epidemiologic concepts.

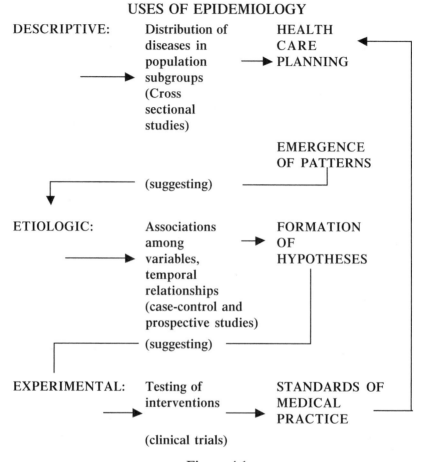

Figure 4.1

4.2   Some Epidemiologic Concepts: Mortality Rates

In 1900, the three major causes of death were influenza or pneumonia, tuberculosis, and gastroenteritis. Today the three major causes of death are heart disease, cancer, and accidents; the fourth is strokes. Stroke deaths have decreased dramatically over the last few decades probably due to the improved control of hypertension, one of the primary risk factors for stroke. These changing patterns of mortality reflect changing environmental conditions, a shift from acute to chronic illness, and an aging population subject to degenerative diseases. We know this from an analysis of *rates*.

The comparison of defined rates among different subgroups of individuals may yield clues to the existence of a health problem and may lead to the specification of conditions under which this identified health problem is likely to appear and flourish.

In using rates, the following points must be remembered:

(1) A rate is a proportion involving a numerator and a denominator.
(2) Both the numerator and the denominator must be clearly defined so that you know to which group (denominator) your rate refers.
(3) The numerator is contained in (is a subset of) the denominator. This is in contrast to a ratio where the numerator refers to a different group from the denominator.

Mortality rates pertain to the number of deaths occurring in a particular population subgroup and often provide one of the first indications of a health problem. The following definitions are necessary before we continue our discussion:

The Crude Annual Mortality Rate (or death rate) is:

$$\frac{\text{the } total\ number \text{ of deaths during a year in the population at risk}}{\text{the population at risk (usually taken as the population at midyear)}}$$

The Cause-Specific Annual Mortality Rate is:

$$\frac{\text{number of deaths occurring due to a } particular\ cause \text{ during the year in the population at risk}}{\text{population at risk (all those alive at midyear)}}$$

The Age-Specific Annual Mortality Rate is:

$$\frac{\text{number of deaths occurring in the given age group during the year in the population at risk}}{\substack{\text{population at risk} \\ \text{(those in that age group alive at midyear)}}}$$

A reason for taking the population at midyear as the denominator is that a population may grow or shrink during the year in question and the midyear population is an estimate of the average number during the year. One can, however, speak of death rates over a five- year period rather than a one-year period, and one can define the population at risk as all those alive at the beginning of the period.

4.3   Age-Adjusted Rates

When comparing death rates between two populations, the age composition of the populations must be taken into account. Since older people have a higher number of deaths per 1,000 people, if a population is heavily weighted by older people, the crude mortality rate would be higher than in a younger population and a comparison between the two groups might just reflect the age discrepancy rather than an intrinsic difference in mortality experience. One way to deal with this problem is to compare age-specific death rates, death rates specific to a particular age group. Another way that is useful when an overall summary figure is required is to use *age-adjusted* rates. These are rates adjusted to what they *would be* if the two populations being compared had the same age distributions as some arbitrarily selected standard population.

For example, the table below shows the crude and age-adjusted mortality rates for the United States at five time periods.15,7 The adjustment is made to the age distribution of the population in 1940 as well as the age distribution of the population in 2000. Thus, we see that in 1991 the age-adjusted rate was 5.1/1000 when adjusted to 1940 standard, but the crude mortality rate was 8.6/1000. This means that if in

1991 the age distribution of the population were the same as it was in 1940, then the death rate would have been only 5.1/1000 people. The crude and age-adjusted rates for 1940 are the same because the 1940 population serves as the "standard" population whose age distribution is used as the basis for adjustment.

When adjusted to the year 2000 standard, the age-adjusted rate was 9.3. If in 1991 the age distribution were the same as in 2000, then the death rate would have been 9.3/1000 people. So age-adjusted rates depend on the standard population being used for the adjustment. Note that the age-adjusted rate based on the population in year 2000 is higher than the age-adjusted rate based on the population in 1940; this is because the population is older in year 2000.

Year	Crude Mortality Rate per 1,000 People	Age-Adjusted Rate (to Population in 1940)	Age-Adjusted Rate (to population in 2000)
1940	10.8	10.8	17.9
1960	9.5	7.6	13.4
1980	8.8	5.9	10.4
1991	8.6	5.1	9.3
2001	8.5	not computed after 1998	8.6

Although both crude and age-adjusted rates have decreased from 1940, the decrease in the age-adjusted rate is much greater. The percent change in crude mortality between 1940 and 1991 was $(10.8 - 8.6)/10.8 = 20.4\%$, whereas the percent change in the age-adjusted rate was $(10.8 - 5.1)/10.8 = .528$ or 52.8%.

The reason for this is that the population is growing older. For instance the proportion of persons 65 years and over doubled between 1920 and 1960, rising from 4.8% of the population in 1920 to 9.6% in 1969. After 1998, the National Center for Health Statistics used the population in 2000 as the standard population against which adjustments were made. The crude rate and the age-adjusted death rate in the year 2001 are similar, and that is because the age distribution in

2001 is similar to the age distribution in 2000 so age-adjustment doesn't really change the mortality rate much.

The age-adjusted rates are fictitious numbers—they do not tell you how many people actually died per 1,000, but how many *would have* died if the age compositions were the same in the two populations. However, they are appropriate for comparison purposes.   Methods to perform age-adjustment are described in Appendix E.

4.4   Incidence and Prevalence Rates

The prevalence rate and the incidence rate are two measures of morbidity (illness).

Prevalence rate of a disease is defined as

$$\frac{\text{Number of persons with a disease}}{\substack{\text{Total number of persons in population} \\ \text{at risk at a particular point in time}}}$$

(This is also known as *point prevalence*, but more generally referred to just as "prevalence.") For example, the prevalence of hypertension in 1973 among black males, ages 30–69, defined as a diastolic blood pressure (DBP) of 95 mm Hg or more at a blood pressure screening program conducted by the Hypertension Detection and Follow-Up Program (HDFP),[16] was calculated to be:

$$\frac{4,268 \; with \; DBP > 95mmHg}{15,190 \; black \; men \; aged \; 30-69 \; screened}$$

$$\times \; 100 = 28.1 \; \text{per} \; 100$$

Several points are to be noted about this definition:
(1) The risk group (denominator) is clearly defined as black men, ages 30–69.
(2) The point in time is specified as time of screening.

(3) The definition of the disease is clearly specified as a diastolic blood pressure of 95 mm Hg or greater. (This may include people who are treated for the disease but whose pressure is still high and those who are untreated.)

(4) The numerator is the subset of individuals in the reference group (denominator) who satisfy the definition of the disease.

The *incidence rate* is defined as:

$$\frac{\text{Number of new cases of a disease per unit of time}}{\text{Total number at risk in beginning of this time period}}$$

For example, studies have found that the ten-year incidence of a major coronary event (such as heart attack) among white men, ages 30–59, with diastolic blood pressure 105 mm Hg or above at the time they were first seen, was found to be 183 per 1,000.[17] This means that among 1,000 white men, ages 30–59, who had diastolic blood pressure above 105 mm Hg at the beginning of the ten-year period of observation, 183 of them had a major coronary event (heart attack or sudden death) during the next ten years. Among white men with diastolic blood pressure of <75 mm Hg, the ten-year incidence of a coronary event was found to be 55/1,000. Thus comparison of these two incidence rates, 183/1,000 for those with high blood pressure versus 55/1,000 for those with low blood pressure, may lead to the inference that elevated blood pressure is a risk factor for coronary disease.

Often one may hear the word "incidence" used when what is really meant is prevalence. You should beware of such incorrect usage. For example, you might hear or even read in a medical journal that the incidence of diabetes in 1973 was 42.6 per 1,000 individuals, ages 45–64, when what is really meant is that the prevalence was 42.6/1,000. The thing to remember is that prevalence refers to the *existence of a disease* at a specific period in time, whereas incidence refers to *new cases* of a disease developing within a specified period of time.

Note that *mortality rate is an incidence rate*, whereas *morbidity may be expressed as an incidence or prevalence rate*. In a chronic disease the prevalence rate is greater than the incidence rate because prevalence includes both new cases and existing cases that may have first occurred a long time ago, but the afflicted patients continued to live with the condition. For a disease that is either rapidly fatal or quickly cured, incidence and prevalence may be similar. Prevalence can be established by doing a survey or a screening of a target population and counting the cases of disease existing at the time of the survey. This is a cross-sectional study. Incidence figures are harder to obtain than prevalence figures since to ascertain incidence one must identify a group of people free of the disease in question (known as a cohort), observe them over a period of time, and determine how many develop the disease over that time period. The implementation of such a process is difficult and costly.

4.5   Standardized Mortality Ratio

The *standardized mortality ratio* (SMR) is the ratio of the number of deaths observed to the number of deaths expected. The number expected for a particular age group for instance, is often obtained from population statistics.

$$SMR = \frac{observed\,deaths}{expected\,deaths}$$

4.6   Person-Years of Observation

Occasionally one sees a rate presented as some number of events *per person-years of observation*, rather than per number of individuals observed during a specified period of time. Per person-years (or months) is useful as a unit of measurement when people are observed for different lengths of time. Suppose you are observing cohorts of people free of heart disease to determine whether the incidence of heart disease is greater for smokers than for those who quit. Quitters need to be defined, for example, as those who quit more than five years prior to the start of observation. One could define quitters differently and get different results, so it is important to specify the definition. Other considerations include controlling for the length of time smoked, which would be a function of age at the start of smoking and age at the start of the observation period, the number of cigarettes smoked, and so forth. But for simplicity, we will assume everyone among the smokers has smoked an equal amount and everyone among the quitters has smoked an equal amount prior to quitting.

We can express the incidence rate of heart disease per some unit of time, say 10 years, as the number who developed the disease during that time, divided by the number of people we observed (number at risk). However, suppose we didn't observe everyone for the same length of time. This could occur because some people moved or died of other causes or were enrolled in the study at different times or for other reasons. In such a case we could use as our denominator the number of *person-years of observation.*

For example, if individual 1 was enrolled at time 0 and was observed for 4 years, then lost to follow-up, he would have contributed 4 person-years of observation. Ten such individuals would contribute 40 person-years of observation. Another individual observed for 8 years would have contributed 8 person-years of observation and 10 such individuals would contribute 80 person-years of observation for a total of 120 person-years. If six cases of heart disease developed among those observed, the rate would be 6 per 120 person-years, rather than 6/10 individuals observed. Note that if the denominator is given as person-years, you don't know if it pertains to 120 people each observed for one year, or 12 people each observed for 10 years or some combination. Another problem with this method of expressing rates is that it reflects the average experience over the time span, but it may be that the rate of heart disease is the same for smokers as for quitters within the first 3 years and the rates begin to separate after that. In any case, various statistical methods are available for use with person-year analysis. An excellent explanation of this topic is given in the book *An Introduction to Epidemiologic Methods*, by Harold A. Kahn.

4.7   Dependent and Independent Variables

In research studies we want to quantify the relationship between one set of variables, which we may think of as predictors or determinants, and some outcome or criterion variable in which we are interested. This outcome variable, which it is our objective to explain, is the dependent variable.

A *dependent variable* is a factor whose value depends on the level of another factor, which is termed an *independent variable*. In the example of cigarette smoking and lung cancer mortality, duration and/or number of cigarettes smoked are independent variables upon which the lung cancer mortality depends (thus, lung cancer mortality is the dependent variable).

4.8   Types of Studies

In Section 1.4 we described different kinds of study designs, in the context of our discussion of the scientific method and of how we know what we know. These were observational studies, which may be cross-sectional, case-control, or prospective, and experimental studies, which are clinical trials. In the following sections we will consider the types of inferences that can be derived from data obtained from these different designs.

The objective is to assess the relationship between some factor of interest (the independent variable), which we will sometimes call exposure, and an outcome variable (the dependent variable).

The observational studies are distinguished by the *point in time when measurements are made on the dependent and independent variables,* as illustrated below. In cross-sectional studies, both the dependent and independent (outcome and exposure) variables are measured at the same time, in the present. In case-control studies, the outcome is measured now and exposure is estimated from the past. In prospective studies, exposure (the independent variable) is measured now and the outcome is measured in the future. In the next section we will discuss the different inferences to be made from cross-sectional versus prospective studies.

	Time of Measurement		
	Past	Present	Future
Cross-Sectional:		exposure outcome	
Case-Control:	exposure	outcome	
Prospective:		exposure	outcome

4.9   Cross-Sectional Versus Longitudinal Looks at Data

Prospective studies are sometimes also known as longitudinal studies, since people are followed longitudinally, over time. Examination of longitudinal data may lead to quite different inferences than those to be obtained from cross-sectional looks at data. For example, consider age and blood pressure.

Cross-sectional studies have repeatedly shown that the average systolic blood pressure is higher in each successive ten-year age group while diastolic pressure increases for age groups up to age 50 and then reaches a plateau. One cannot, from these types of studies, say that blood pressure rises with age because the pressures measured for 30-year-old men, for example, were not obtained on the same individuals ten years later when they were 40, but were obtained for a different set of 40-year-olds. To determine the effect of age on blood pressure we would need to take a longitudinal or prospective look at the same individuals as they get older. One interpretation of the curve observed for diastolic blood pressure, for instance, might be that individuals over 60 who had very high diastolic pressures died off, leaving only those individuals with lower pressure alive long enough to be included in the sample of those having their blood pressure measured in the cross-sectional look.

The diagrams in Figure 4.2 illustrate the possible impact of a "cohort effect," a cross-sectional view and a longitudinal view of the same data. (Letters indicate groups of individuals examined in a particular year.)

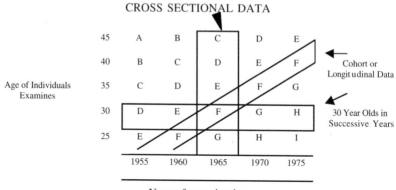

Figure 4.2

If you take the blood pressure of all groups in 1965 and compare group F to group D, you will have a cross-sectional comparison of 30-year-olds with 40-year-olds at a given point in time. If you compare group F in 1965 with group F (same individuals) in 1975, you will have a longitudinal comparison. If you compare group F in 1965 with group H in 1975, you will have a comparison of blood pressures of 30-year-olds at different points in time (a horizontal look).

These comparisons can lead to quite different conclusions, as is shown by the schematic examples in Figure 4.3 using fictitious numbers to represent average diastolic blood pressure.

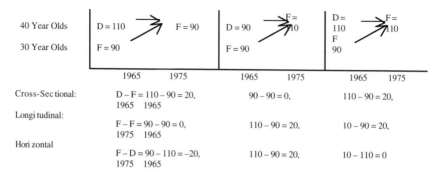

Figure 4.3

In example (1) measurements in 1965 indicate that average diastolic blood pressure for 30-year-olds (group F) was 90 mm Hg and
for 40-year-olds (group D), it was 110 mm Hg. Looking at group F ten
years later, when they were 40-year-olds, indicates their mean diastolic
blood pressure was 90 mm Hg. The following calculations result:

cross-sectional look: D   –   F = 110 – 90 = 20
                            1965   1965

conclusion: 40-year-olds have higher blood pressure than
                            30-year-olds (by 20 mm Hg).

longitudinal look: F   –   F = 90 – 90 = 0
                            1975   1965

conclusion: Blood pressure does not rise with age.

horizontal look: F   –   D = 90 – 110 = –20
(cohort comparisons)   1975   1965

conclusion: 40-year-olds in 1975 have lower blood pres-
                            sure than 40-year-olds did in 1965.

A possible interpretation: Blood pressure does not rise with age, but
different environmental forces were operating for the F cohort, than
for the D cohort.

In example (2) we have

cross-sectional look: D   –   F = 90 – 90 = 0 mm Hg
                            1965   1965

conclusion: From cross-sectional data, we conclude that
                            blood pressure is not higher with older age.

longitudinal look: F   –   F = 110 – 90 = 20
                              1975   1965

conclusion: From longitudinal data we conclude that
                             blood pressure goes up with age.

horizontal look: F   –   D = 110 – 90 = 20
                              1975   1965

conclusion: 40-year-olds in 1975 have higher blood pres-
                             sure than 40-year-olds in 1965.

A possible interpretation: Blood pressure does rise with age and different environmental factors operated on the F cohort than on the D cohort.

   In example (3) we have

cross-sectional look: D   –   F = 110 – 90 = 20
                              1965   1965

conclusion: Cross-sectionally, there was an increase in
                             blood pressure for 40-year-olds over that for
                             30-year-olds.

longitudinal look: F   –   F = 110 – 90 = 20
                              1975   1965

conclusion: Longitudinally it is seen that blood pressure
                             increases with increasing age.

horizontal look: F   –   D = 110 – 110 = 0
                              1975   1965

conclusion: There was no change in blood pressure
                                   among 40-year-olds over the ten-year period.

A possible interpretation: Blood pressure does go up with age (supported by both longitudinal and cross-sectional data) and environmental factors affect both cohorts similarly.

4.10 Measures of Relative Risk:
Inferences From Prospective Studies:
the Framingham Study

In epidemiologic studies we are often interested in knowing how much more likely an individual is to develop a disease if he or she is exposed to a particular factor than the individual who is not so exposed. A simple measure of such likelihood is called *relative risk (RR)*. It is the ratio of two incidence rates: *the rate of development of the disease for people with the exposure factor, divided by the rate of development of the disease for people without the exposure factor*. Suppose we wish to determine the effect of high blood pressure (hypertension) on the development of cardiovascular disease (CVD). To obtain the relative risk we need to calculate the incidence rates. We can use the data from a classic prospective study, the Framingham Heart Study.[18]

This was a pioneering prospective epidemiologic study of a population sample in the small town of Framingham, Massachusetts. Beginning in 1948 a *cohort* of people was selected to be followed up biennially. The term *cohort* refers to a group of individuals followed longitudinally over a period of time. A birth cohort, for example, would be the population of individuals born in a given year. The Framingham cohort was a sample of people chosen at the beginning of the study period and included men and women aged 30 to 62 years at the start of the study. These individuals were observed over a 20-year period, and morbidity and mortality associated with cardiovascular disease were determined. A standardized hospital record and death certificate were obtained, clinic examination was repeated at two-year intervals, and the major concern of the Framingham study has been to evaluate the relationship of characteristics determined in *well* persons to the subsequent development of disease.

Through this study "risk factors" for cardiovascular disease were identified. The *risk factors* are antecedent physiologic characteristics or dietary and living habits, whose presence increases the individual's probability of developing cardiovascular disease at some future time. Among the most important predictive factors identified in the Framingham study were *elevated blood pressure, elevated serum cholesterol, and cigarette smoking.* Elevated blood glucose and abnormal resting electrocardiogram findings are also predictive of future cardiovascular disease.

Relative risk can be determined by the following calculation:

$$\frac{\text{incidence rate of cardiovascular disease (new cases)}}{\text{incidence rate of CVD in the given time period among}}$$

incidence rate of cardiovascular disease (new cases)
over a specified period of time among people free
of CVD at beginning of the study period who have
the risk factor in question (e.g., high blood pressure)
_____
incidence rate of CVD in the given time period among
people free of CVD initially, who do not have the risk
factor in question (normal blood pressure)

From the Framingham data we calculate for men in the study the

$$\begin{array}{c} \text{RR of CVD within} \\ \text{18 years after first} \\ \text{exam} \end{array} = \frac{\begin{array}{c} 353.2/10,000 \text{ persons at risk} \\ \text{with definite hypertension} \end{array}}{\begin{array}{c} 123.9/10,000 \text{ persons at risk} \\ \text{with no hypertension} \end{array}}$$

$$\frac{353.2}{123.9} = 2.85$$

This means that a man with definite hypertension is 2.85 times more likely to develop CVD in an 18-year period than a man who does not have hypertension. For women the relative risk is

$$\frac{187.9}{57.3} = 3.28$$

This means that hypertension carries a somewhat greater relative risk for women. But note that the *absolute* risk for persons with definite hypertension (i.e., the incidence of CVD) is greater for men than for women, being 353.2 per 10,000 men versus 187.9 per 10,000 women.

The incidence rates given above have been age-adjusted. Age adjustment is discussed in Section 4.3. Often one may want to adjust for other variables such as smoking status, diabetes, cholesterol levels, and other factors that may also be related to outcome. This may be accomplished by multiple logistic regression analysis or by Cox proportional hazards analysis, which are described in Sections 4.16 and 4.18, respectively, but first we will describe how relative risk can be calculated from prospective studies or estimated from case-control studies.

4.11 Calculation of Relative Risk from Prospective Studies

Relative risk can be determined directly from prospective studies by constructing a 2 × 2 table as follows:[19]

DISEASE
(developing in the specified period)

		Yes	No	
RISK FACTOR (determined at beginning of study period)	PRESENT (high blood pressure)	$a = 90$	$b = 403$	$a + b = 493$ (persons with factor)
	ABSENT (normal blood pressure)	$c = 70$	$d = 1201$	$c + d = 1271$ (persons without factor)

Relative risk is

$$\frac{incidence\ of\ disease\ among\ those\ with\ high\ BP}{incidence\ of\ disease\ among\ those\ with\ normal\ BP} =$$

$$\frac{a/(a+b)}{c/(c+d)} = \frac{90/493}{70/1271} = 3.31$$

Relative risk, or hazard ratio, can be calculated from Cox proportional hazards regression models (which allow for adjustment for other variables) as described in Section 4.19.

4.12 Odds Ratio: Estimate of Relative Risk from Case-Control Studies

A case-control study is one in which the investigator seeks to establish an association between the presence of a characteristic (a risk factor) and the occurrence of a disease *by starting out with a sample of diseased persons and a control group of nondiseased persons and by noting the presence or absence of the characteristic in each of these two groups.* It can be illustrated by constructing a 2×2 table as follows:

		DISEASE	
		PRESENT	ABSENT
RISK FACTOR	PRESENT	a	b
	ABSENT	c	d
		$a + c$ (number of persons *with* disease)	$b + d$ (number of persons *without* disease)

The objective is to determine if the proportion of persons with the disease who have the factor is greater than the proportion of persons without the disease who have the factor. In other words, it is desired to know whether $a/(a + c)$ is greater than $b/(b + d)$.

Case-control studies are often referred to as *retrospective studies* because the investigator must gather data on the *independent* variables *retrospectively*. The dependent variable—the presence of disease—is obtained at time of sampling, in contrast to prospective studies where the independent variables are measured at time of sampling and the dependent variable is measured at some future time (i.e., when the disease develops). The real distinction between case-control or retrospective studies and prospective studies has to do with selecting individuals for the study—those *with and without the disease* in case-control/retrospective studies, and those *with and without the factor* of interest in prospective studies.

Since in prospective studies *we sample the people with the characteristic of interest and the people without the characteristic,* we can obtain the relative risk *directly* by calculating the incidence rates of disease in these two groups. In case-control studies, however, *we sample patients with and without the disease,* and note the presence or absence of the characteristic of interest in these two groups; we do not have a direct measure of *incidence* of disease. Nevertheless, making certain assumptions, we can make some approximations to what the relative risk would be if we could measure incidence rates through a prospective study. These approximations hold best for diseases of *low incidence.*

To estimate relative risk from case-control studies note that

$$\frac{a/(a+b)}{c/(c+d)} = \frac{a(c+d)}{c(a+b)}$$

Now assume that since the disease is not all that common, c is negligible in relation to d (in other words among people without the risk factor there aren't all that many who will get the disease, relative to the number of people who will not get it). Assume also that, *in the population, a* is negligible relative to b, since even among people with the risk factor not all that many will get the disease in comparison to the number who won't get it. Then the terms in the parentheses become d in the numerator and b in the denominator so that

$$\frac{a(c+d)}{c(a+b)} \quad reduces\ to \quad OR = \frac{ad}{bc}$$

This is known as the *odds ratio* (OR) and is a good estimate of relative risk when the disease is rare.

An example of how the odds ratio is calculated is shown below. In a case-control study of lung cancer the table below was obtained.[20] Note that we are not sampling smokers and nonsmokers here. Rather we are sampling those with and without the disease. So although in the *population at large a* is small relative to *b*, in this sample it is not so.

	Patients with Lung Cancer		Matched Controls with Other Diseases	
Smokers of 15–24 cigarettes daily	475	*a*	431	*b*
Nonsmokers	7	*c*	61	*d*
	(persons with disease)		(persons without disease)	

The odds ratio, as an estimate of the relative risk of developing lung cancer for people who smoke 15–24 cigarettes a day, compared with nonsmokers is

$$Odds\ ratio = \frac{475 \times 61}{431 \times 7} = 9.60 = estimate\ of\ relative\ risk$$

This means that smokers of 15–24 cigarettes daily are 9.6 times more likely to get lung cancer than are nonsmokers.

One more thing about the odds ratio: it is the ratio of odds of lung cancer for those who smoke 15–24 cigarettes a day, relative to odds of lung cancer for those who don't smoke. In the example above, we get

$$for\ smokers: odds\ of\ lung\ cancer\ are\ \frac{475}{431}$$

$$for\ nonsmokers: odds\ of\ lung\ cancer\ are\ \frac{7}{61}$$

$$ratio\ of\ odds = \frac{475/431}{7/61}$$

So the point is, the odds ratio is the odds ratio, whether the disease is rare or not. It is always the ratio of odds of disease for those with the exposure versus the odds of disease for those without the exposure. But when the disease is rare, it is also a good estimate of the relative risk. We can also put confidence limits on the odds ratio, shown in Appendix F. Odds ratios can be calculated from logistic regression (which allow for adjustment for other variables) as described in Section 4.17.

4.13 Attributable Risk

Attributable risk (AR) is:
          risk in exposed—risk in unexposed individuals.

Population attributable risk (PAR) is:
          AR x risk factor prevalence

   While relative risk pertains to the risk of a disease in exposed persons *relative to the risk* in the unexposed, the attributable risk pertains to *the difference in absolute risk* of the exposed compared to the unexposed persons. It may tell us how much *excess risk* there is due to the exposure in the exposed. In the example in Section 4.11, the 10 year risk among those with high blood pressure was 90/493 = 0.183, (or 183 per 1000 people with high blood pressure) while in those with normal blood pressure it was 70/1271 = 0.055 (or 55 per 1000 with normal pressure).

Thus the attributable risk in those exposed (i.e. with high blood pressure) is 0.183-0.055 = 0.128 (128 per 1000). In other words, heart disease events in 128 of the 183 people per 1000 with high blood pressure can be attributed to the high blood pressure. We can also express this excess as a percentage of the risk in the exposed that is attributable to the exposure:

$$\frac{128/1000}{183/1000} = 128 = .70 \; or \; 70\%$$

BUT, we must be very careful about such attribution—it is only valid when we can assume the exposure is causes the disease (after taking into account confounding and other sources of bias).

Population attributable risk (PAR) is a useful measure when we want to see how we could reduce morbidity or mortality by eliminating a risk factor. It depends on the prevalence of the risk factor in the population as noted above. Here is an example from the Women's Health Initiative (described in more detail in Chapter 6). It was found in a clinical trial that postmenopausal women who were taking estrogen plus progestin had an annualized rate of coronary heart disease of 39 per 10,000 compared to a rate of 33 per 10,000 for women taking placebo.[21]

Thus:

$$AR = \frac{39 - 33}{10,000} = .0006$$

or 6 excess coronary heart disease events per 10,000 women taking this preparation, per year.

Since approximately 6,000,000 women were taking that hormone preparation at the time (exposed), then .0007 x 6,000,000 = 3600 coronary heart disease events per year could be *attributed* to taking estrogen plus progestin.

The prevalence of use of estrogen plus progestin estimated from the same study when it was first begun, was about 18%. If we use this estimate,

PAR = AR x prevalence of risk factor = .0006 x .18 = .000108;

Thus, if use of estrogen plus progestin were eliminated there would be 10.8 per 100,000 postmenopausal women who had fewer coronary heart disease events.

4.14 Response Bias

There are many different types of bias that might lead you to either underestimate or overestimate the size of a relative risk of odds ratio, and it is important to try to anticipate potential sources of bias and avoid them. The illustration on the next page shows the impact of one kind of possible bias: *ascertainment or response bias.*

Assume that you have the following situation. Of 100 people exposed to a risk factor, 20% develop the disease and of a 100 people unexposed, 16% develop the disease, yielding a relative risk of 1.25, as shown in the illustration.

Now imagine that only 60% of the exposed respond to follow-up, or are ascertained as having or not having the disease, *a 60% response rate among the exposed.* Assume further that all of the ones who don't respond happen to be among the ones who *don't* develop disease. The relative risk would be calculated as 2.06.

Now imagine that only 60% of the nonexposed reply, *a 60% response rate among the nonexposed*, and all of the nonexposed who don't respond happen to be among the ones who don't have the disease. Now the relative risk estimate is 0.75.

To summarize, you can get conflicting estimates of the relative risk if you have differential response rates. Therefore, it is very important to get as complete a response or ascertainment as possible. The tables showing these calculations follow.

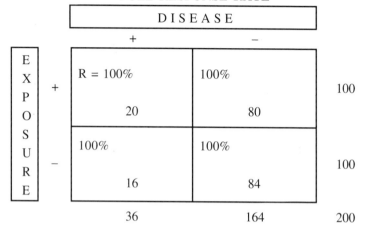

FULL RESPONSE RATE

	DISEASE		
	+	−	
EXPOSURE +	R = 100% 20	100% 80	100
EXPOSURE −	100% 16	100% 84	100
	36	164	200

$$RR = \frac{20/100}{16/100} = \frac{.20}{.16} = 1.25$$

DIFFERENTIAL RESPONSE RATE

	DISEASE		
	+	−	
EXPOSURE +	R = 100% 20	50% 40	60 (response rate = 60%)
EXPOSURE −	100% 16	100% 84	100
	36	124	160

$$RR = \frac{20/60}{16/100} = \frac{.33}{.16} = 2.06$$

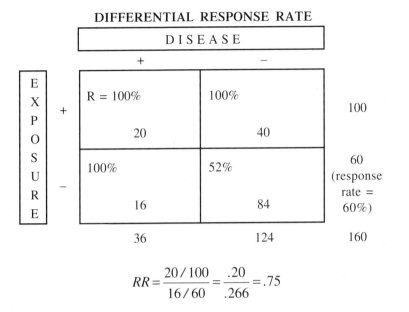

$$RR = \frac{20/100}{16/60} = \frac{.20}{.266} = .75$$

4.15 Confounding Variables

A *confounding variable* is one that is closely associated with both the independent variable and the outcome of interest in those unexposed. For example, a confounding variable in studies of coffee and heart disease may be smoking. Since some coffee drinkers are also smokers, if a study found a relationship between coffee drinking (the independent variable) and development of heart disease (the dependent variable), it could really mean that it is the smoking that causes heart disease, rather than the coffee. In this example, smoking is the confounding variable.

If *both* the confounding variable and the independent variable of interest are closely associated with the dependent variable, then the observed relationship between the independent and dependent variables may really be a reflection of the *true* relationship between the confounding variable and the dependent variable. An intuitive way to look at this is to imagine that if a confounding variable were perfectly associated with an independent variable, it could be substituted for it. It is important to account or adjust for confounding variables *in the design and statistical analysis of studies* in order to avoid wrong inferences.

There are several approaches to dealing with potential confounders. One approach is to deal with it in the study design by matching, for example, as described in Section 4.16 below; another way of controlling for confounding variables is in the data analysis phase, by using multivariate analysis, as described in Sections 4.17, 4.18 and 4.20 below. An excellent discussion is found in *Modern Epidemiology* by Kenneth J. Rothman and Sander Greenland.

4.16 Matching

One approach to dealing with potential confounders is to match subjects in the two groups on the confounding variable. In the example discussed above concerning studies of coffee and heart disease, we might match subjects on their smoking history, since smoking may be a confounder of the relationship between coffee and heart disease. Whenever we enrolled a coffee drinker into the study, we would determine if that person was a smoker. If the patient was a smoker, the next patient who would be enrolled who was not a coffee drinker (i.e., a member of the comparison group), would also have to be a smoker. For each coffee-drinking nonsmoker, a non–coffee-drinking nonsmoker would be enrolled. In this way we would have the same number of smokers in the two groups. This is known as *one-to-one matching*. There are other ways to match and these are discussed more fully in the Suggested Readings section, especially in *Statistical Methods for Comparative Studies* by Anderson et al. and in *Causal Relationships in Medicine* by J. Mark Elwood.

In case-control studies finding an appropriate comparison group may be difficult. For example, suppose an investigator is studying the effect of coffee on pancreatic cancer. The investigator chooses as controls, patients in the hospital at the same time and in the same ward as the cases, but with a diagnosis other than cancer. It is possible that patients hospitalized for gastrointestinal problems other than cancer might have voluntarily given up coffee drinking because it bothered their stomachs. In such a situation, the coffee drinking habits of the two groups might be similar and the investigator might not find a greater association of coffee drinking with cases than with controls. A more appropriate group might be patients in a different ward, say an orthopedic ward. But here one would have to be careful to match on age, since orthopedic patients may be younger than the other cases if the hospital happens to be in a ski area, for example, where reckless skiing leads to broken legs, or they may be substantially older than the other cases if there are many patients with hip replacements due to falls in the elderly, or osteoarthritis.

It needs to be pointed out that the factor that is matched cannot be evaluated in terms of its relationship to outcome. Thus, if we are comparing two groups of women for the effect of vitamin A intake on cervical cancer and we do a case-control study in which we enroll cases of cervical cancer and controls matched on age, we will not be able to say from this study whether age is related to cervical cancer. This is because we have ensured that the age distributions are the same in both the case and control groups by matching on age, so obviously we will not be able to find differences in age between the groups.

Some statisticians believe that matching is often done unnecessarily and that if you have a large enough study, simple randomization or stratified randomization is adequate to ensure a balanced distribution of confounding factors. Furthermore, multivariate analysis methods, such as logistic regression or proportional hazards models, provide another, usually better, way to control for confounders. A good discussion of matching can be found in the book *Methods in Observational Epidemiology*, by Kelsey, Thompson, and Evans.

4.17 Multiple Logistic Regression

Multiple logistic regression analysis is used to calculate the probability of an event happening as a function of several independent variables. It is useful in controlling for confounders when examining the relationship between an independent variable and the occurrence of an outcome (e.g., such as heart attack) within a specified period of time. The equation takes the form of

$$P(event) = \frac{1}{1 + e^{-k}}$$

$$where\ k = C_0 + C_1 X_1 + C_2 X_2 + C_3 X_3 + ... + C_m X_m$$

Each X_i is a particular independent variable and the corresponding coefficients, C's, are calculated from the data obtained in the study. For example, let us take the Framingham data for the probability of a man developing cardiovascular disease within 8 years. Cardiovascular disease (CVD) was defined as coronary heart disease, brain infarction, intermittent claudication, or congestive heart failure.

$$P(CVD) =$$

$$\frac{1}{1 + e^{-[-19.77+.37(age)-.002(age)^2+.026(chl)+.016(SBP)+.558(SM)+1.053(LVH)+.602(Gl)-.00036(chl \times age)]}}$$

where chl = serum cholesterol,
  SBP = systolic blood pressure,
  SM = 1 if yes for smoking, 0 if no,
  LVH = left ventricular hypertrophy, 1 if yes, 0 if no,
  Gl = glucose intolerance, 1 if yes, 0 if no.

For example, suppose we consider a 50-year-old male whose cholesterol is 200, systolic blood pressure is 160, who smokes, has no LVH, and no glucose intolerance. When we multiply the coefficients by this individual's values on the independent variables and do the necessary calculations we come up with a probability of .17. This means that this individual has 17 chances in a 100 of developing some form of cardiovascular disease within the next 8 years.

The coefficients from a multiple logistic regression analysis can be used to calculate the odds ratio for one factor while controlling for all the other factors. The way to do this is to take the natural log e raised to the coefficient for the variable of interest, if the variable is a dichotomous one (i.e., coded as 1 or 0). For example, the odds of cardiovascular disease for smokers relative to nonsmokers among males, while controlling for age, cholesterol, systolic blood pressure, left ventricular hypertrophy, and glucose intolerance is $e^{.558} = 1.75$. This means that a person who smokes has 1.75 times higher risk of getting CVD (within 8 years) than the one who doesn't smoke if these two individuals are equal with respect to the other variables in the equation. This is equivalent to saying that the smoker's risk is 75% higher than the nonsmoker's.

If we want to compare the odds of someone with a systolic blood pressure of 200 versus someone with systolic blood pressure of 120, all other factors being equal, we calculate it as follows:

$$OR = e^{\beta(200-120)} = e^{.016(80)} = e^{1.28} = 3.6$$

The man with systolic blood pressure of 200 mm Hg is 3.6 times more likely to develop disease than the one with pressure of 120. (Of course, this would imply that someone with systolic blood pressure of 260 would also be 3.6 times more likely to develop CVD than one with pressure of 180. If the assumption of a linear increase in risk didn't hold, then the prediction would be incorrect.)

Logistic regression can also be used for case-control studies. In this case raising e to the coefficient of the variable of interest also gives us the odds ratio, but we cannot use the equation to predict the probability of an event, since we have sampled from cases and controls, not from a general population of interest.

Multiple logistic regression analysis has become widely used largely due to the advent of high-speed computers, since calculating the coefficients requires a great deal of computer power. Statistical packages are available for personal computers.

Multiple logistic regression is appropriate when the dependent variable (outcome) is dichotomous (i.e., can be coded as 1 = event, 0 = no event), and when the question deals with the occurrence of the event of interest within a specified period time and the people are all followed for that length of time. However, when follow-up time for people in the study differs, then survival analysis should be used, as described in sections 4.19 and 4.20.

4.18 Confounding By Indication

One type of confounding that can occur in observational studies when you are looking at the effects of drug treatment on future events, is *confounding by indication*. For example suppose you want to compare the effects on heart disease of different drugs for high blood pressure. You determine what antihypertensive drugs study participants are taking at a baseline examination and then you follow them forward in time to see who develops heart disease. The problem may be that the reason people were taking different drugs at baseline was that they had different indications for them and the doctors prescribed medications appropriate to those indications. Thus people with kidney disease may have been prescribed a different drug to control high blood pressure than those with angina, or than those with no other medical conditions, and each of those indications may be differently related to the outcome of heart disease. Only a clinical trial, where the patients are randomly assigned to each drug treatment, can truly answer the question about different effects of the drugs.

However, there are ways to minimize the confounding by indication in observational studies; one way is to exclude from the analysis people who have angina or kidney disease in the above example. Another method gaining in use is *propensity analysis*.[22] The general idea is that you predict who is likely to be taking the drug from the independent variables you have measured and calculate an index of "propensity" for taking the drug. That propensity score is then entered as an independent variable in your final multivariate equation, along with a subset of the variables that you are controlling for. Each person's data then, include the values of the covariates and his/her propensity score.

For example in the hypertension example, if we are looking to see whether a calcium channel blocker is associated with mortality, we want to take into account that in an observational study people might be more likely to have had a calcium channel blocker prescribed if they had angina for example, and we know that angina is related to mortality. We might then take the following steps:

(1) Calculate a multiple logistic regression where:
Y=1 if on drug, 0 otherwise (dependent variable)
X's (independent variables) = age, race/ethnicity, angina, BMI, systolic blood pressure, other covariates that might influence prescribing a calcium channel blocker.

(2) Calculate a propensity score for each person (an index based on the regression developed above)

(3) Calculate the regression you are really interested in which is to determine the association of calcium channel blockers with mortality after controlling for potential confounders, where:
Z=1, if mortal event, 0 otherwise (dependent variable)
X's = propensity score, age, race/ethnicity, plus some of the other relevant covariates that were in the original propensity equation.

One problem with propensity analysis is that it depends on which variables you have to enter into the equation to get the propensity score; they may describe propensity, or they may not. There are differing views on how useful such an analysis is. It is described in more detail in the references given at the end.

4.19 Survival Analysis: Life Table Methods

Survival analysis of data should be used when the follow-up times differ widely for different people or when they enter the study at different times. It can get rather complex and this section is intended only to introduce the concepts. Suppose you want to compare the survival of patients treated by two different methods and suppose you have the data shown below.[23] We will analyze it by using the Kaplan–Meier survival curves.

DEATHS AT A GIVEN MONTH IN TWO GROUPS

Status: (D = dead at that month; L = living at that month).

(The + means patient was lost to follow-up and last seen alive at that month.)

Status: D L D D D   D D D D L
Group A: 4, 5+, 9, 11, 12   Group B: 2, 3, 4, 5, 6+

In each group four patients had died by 12 months, and one was seen alive some time during that year, so we don't know whether that patient was dead or alive at the end of the year. If we looked at the data in this way, we would have to say that the survival by one year was the same in both groups.

	Group	
At end of 12 months	**A**	**B**
Dead	4	4
Alive	1	1
Survival rate	20%	20%

However, a more appropriate way to analyze such data is through *survival curves*. The points for the curves are calculated as shown in the table below.

Col.1	Col.2	Col.3	Col.4	Col.5	Col.6	Col.7
Case#	Time in Mos.	Status	# Pts. Enter	Prop. Dead $q_i=$ $=\frac{dead}{entered}$	Prop. Surv. $P_1=$ $1-q_i$	Cum. Surv. $P_1=$ $P_{i-1} \times P_i$
Group A						
1	4	dead	5	1/5 = 0.2	0.80	1 x .8 = .8
2	5	surv	4	0/4 = 0.0	1.00	.8 x 1 = .8
3	9	dead	3	1/3 = 0.33	0.67	.8 x .67 = .53
4	11	dead	2	1/2 = 0.5	0.50	.53 x .5 = 27
5	12	dead	1	1/1 = 1.0	0.00	.27 x 0 = 0

Col.1	Col.2	Col.3	Col.4	Col.5	Col.6	Col.7
Case#	Time in Mos.	Status	# Pts. Enter	Prop. Dead $q_i=$ $=\dfrac{\text{dead}}{\text{entered}}$	Prop. Surv. $P_i=$ $1-q_i$	Cum. Surv. $P_i=$ $P_{i-1} \times P_i$
Group B						
1	2	dead	5	1/5 = 0.2	0.80	$1 \times .8 = .8$
2	3	dead	4	1/4 = 0.25	0.75	$.8 \times .75 = .6$
3	4	dead	3	1/3 = 0.33	0.67	$.6 \times .67 = .4$
4	5	dead	2	1/2 = 0.5	0.50	$.4 \times .5 = .2$
5	6	surv	1	0/1 = 0.0	1.00	$.2 \times .1 = .2$

First of all, the patients are placed in order of the time of their death or the last time they were seen alive. Let us go through the third row for group A. as an example.

The third patient died at 9 months (columns 1 and 2). At the beginning of the 9th month there were three patients at risk of dying(out of the total of five patients who entered the study). This is because one of the five patients had already died in the 4th month (case #1),

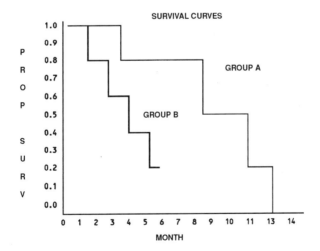

Figure 4.4

and one was last seen alive at the 5th month (case #2) and so wasn't available to be observed. Out of these three patients at risk in the beginning of the 9th month, one died (case #3). So the probability of dying in the 9th month is 1/3 and we call this q_i where i in this case refers to the 9th month. Therefore the proportion surviving in the 9th month is $p_i = 1 - q_i = 1 - .33 = .67$.

The cumulative proportion surviving means the proportion surviving up through the 9th month. To survive through the 9th month, a patient had to have survived to the end of month 8 *and* have survived in month 9. Thus, it is equal to the cumulative probability of surviving *up to* the 9th month, which is .8, from column 7 row 2, *and* surviving in the 9th month, which is .67. We multiply these probabilities to get .8 × .67 = .53 as the probability of surviving through the 9th month. If we plot these points as in Figure 4.4, we note that the two survival curves look quite different and that group A did a lot better.

Survival analysis gets more complicated when we assume that patients who have been lost to follow-up in a given interval of time would have died at the same rate as those patient on whom we had information. Alternatively, we can make the calculations by assuming they all died within the interval in which they were lost to follow-up, or they all survived during that interval.

Survival analysis can also be done while controlling for confounding variables, using the Cox proportional hazards model.

4.20 Cox Proportional Hazards Model

The Cox proportional hazards model is a form of multivariate survival analysis that can control for other factors. *The dependent variable is time to event (or survival time),* which could be death, heart attack, or any other event of interest. This is in contrast to multiple logistic regression, where the dependent variable is a yes-or-no variable.

Cox proportional hazards model is appropriately used when there are different follow-up times because some people have withdrawn from the study, or can't be contacted. People falling into one of those categories are considered to have "censored" observations. If the event of interest is, say, stroke, then people who died during the study from accidental causes, would also be "censored" because we couldn't know whether they would have gone on to have a stroke or not, had they lived to the end of the study.

The coefficients from this analysis can be used to calculate the relative risk of event, after controlling for the other covariates in the equation. The relative risk from Cox proportional hazards models is more accurately called the "hazard ratio" but will be referred to as relative risk here for the sake of simplifying the explanations. An example of how to interpret results from such an analysis is given from the Systolic Hypertension in the Elderly Program (SHEP). This was a study of 4,736 persons over age 60 with isolated systolic hypertension (i.e., people with high systolic blood pressure and normal diastolic blood pressure) to see if treatment with a low-dose diuretic and/or beta-blocker would reduce the rate of strokes compared with the rate in the control group treated with placebo.

A sample of a partial computer printout of a Cox regression analysis from the SHEP study is shown below. The event of interest is stroke in the placebo group.

Independent Variable	Beta Coefficient	S.E.	e^{Beta} = RR
Race	−0.1031	.26070	0.90
Sex (male)	0.1707	.19520	1.19
Age	0.0598	.01405	1.06
History of diabetes	0.5322	.23970	1.70
Smoking (Baseline)	0.6214	.23900	1.86

Let us look at the history of diabetes. The RR = $e^{.5322}$ = 1.70, which is the natural logarithm e raised to the power specified by the beta coefficient; e = 2.7183. (Don't ask why.) This means that a person with untreated systolic hypertension who has a history of diabetes has 1.7 times the risk of having a stroke than a person with the same other characteristics but no diabetes. This can also be stated as a 70% greater risk. The 95% confidence limits for the relative risk are 1.06, 2.72, meaning that we are 95% confident that the relative risk of stroke for those with a history of diabetes lies within the interval between 1.06, 2.72. The formula for the 95% confidence interval for relative risk is

$$\text{Limit } 1 = e^{[\text{beta} - 1.96 \text{ (S.E.)}]}$$

$$\text{Limit } 2 = e^{[\text{beta} + 1.96 \text{ (S.E.)}]}$$

If we are dealing with a continuous variable, like age, the RR is given per one unit or 1 year age increase. The relative risk per 5-year increase in age is

$$e^{5 \text{ beta}} = e^{5 \times .0598} = 1.35$$

There is a 34% increase in risk of future stroke per 5-year greater age at baseline, controlling for all the other variables in the model. To calculate confidence intervals for this example, you also need to multiply the s.e. by 5 (as well as multiplying the beta by 5), so the 95% confidence intervals of RR are: [1.18, 1.55] for a 5-year increase in age.

Appendix G provides additional information on exploring a J or U shape relationship between a variable and the outcome.

4.21 Selecting Variables For Multivariate Models

Suppose we want to determine the effect of depression on subsequent heart disease events using data from a prospective follow-up study. We can run Cox proportional hazards models to obtain relative risk of depression for heart disease endpoints, but we want to control for confounders. Otherwise any association we see might really be a reflection of some other variable, like say smoking which is related to depression and is also a risk factor for heart disease. How shall we go about deciding which variables to put in the model? There is no single answer to that question and different experts hold somewhat different views, although it is generally agreed that known confounders should be included. So we would put in variables that are significantly related to depression and also to heart disease among the non-exposed, i.e. non-depressed. We would not include variables that are presumed from past experience to be either highly correlated to depression (referred to as collinear) or intermediate in the pathway relating depression to heart disease, such as say some blood biomarker related to heart disease which is elevated by depression. In such a case, the elevation in the blood biomarker is intermediate between depression and heart disease; it may be the first manifestation of heart disease. The point is that a lot of judgment has to be used in selecting variables for inclusion. The objective is to see whether effects of depression that were found remain after accounting for other established risk factors.

One strategy is to start by getting the relative risk of depression alone, and then add successively, one at a time, other potential confounders to see if they change the relative risk for depression by 10% or more (though that is an arbitrary percentage). Variables that qualify by this criterion are kept in the model. For example, in the study of depression and deaths from cardiovascular causes, among post-menopausal women enrolled in the Women's Health Initiative, who had no prior cardiovascular disease, the relative risk associated with depression controlling for age and race was 1.58; adding education and income to that resulted in a relative risk of 1.52. Adding additional variables to the model (diabetes, hypertension, smoking, high cholesterol requiring pills, hormone use, body mass index and physical activity) didn't change things, resulting in a relative risk of 1.50. So it was concluded that depression was an independent risk factor for cardiovascular death.

Now if one were interested in developing a model that would predict risk (rather than one that would evaluate whether a particular risk factor was an independent contributor to risk, as in the example above), one might chose other strategies, like stepwise regression. Stepwise regression can be forward stepwise or backward stepwise and computer programs calculating regressions ask you to specify which you want.

The basic principle is that in forward stepwise regression you first enter the single variable that has the highest correlation with your outcome, then keep adding variables one at a time until you add one that is not statistically significant at some pre-chosen level, then stop. In backward stepwise regression you start out with all the potential variables that can be explanatory and drop them one at a time, eliminating the one that is least significant (has the highest p value) first, until dropping the next variable would result in a poorer model.

Many people don't like stepwise regression because it is somewhat arbitrary; it depends on the significance levels you chose to enter or leave the model and also a variable may have quite a different effect if it is in a model with some other variables that might modify it, rather than when it is in the model alone. Another strategy is to look at all possible regressions—i.e. look at all two variable models, then at all possible 3 variable models and so on. You select the best one according to how much of the variance in the dependent variable is explained by the model. An excellent discussion of variable selection in epidemiologic models is by Sander Greenland[24] and also in the advanced texts noted in the Suggested Readings section.

4.22 Interactions: Additive and Multiplicative Models

An interaction between two variables means that the effect of one variable on the outcome of interest is different depending on the level of the other variable, as described in Section 3.26. Interactions may be *additive*, where the joint effect of two variables is *greater than the sum of their individual effects*, or *multiplicative,* where the joint effect of the two variables *is greater than the product of the individual effects* of each variable.

Logistic and Cox regression models are inherently multiplicative. When we say that smoking carries a relative risk of 2 for coronary heart disease for example, we mean that smokers are two *times* more likely than non-smokers to get the disease. We may want to know if there is an interaction between smoking and hypertension. In other words, we want know whether smoking among hypertensives has a greater effect on heart attacks than we would expect from knowing the separate risks of smoking and hypertension. We can test the hypothesis of no interaction versus the alternative hypothesis of an interaction, but first we need to know what we would expect under a multiplicative model if there were no interaction.

Consider two dichotomous variable A and B.

The table below shows the pattern of relative risks expected under the multiplicative model if there is no multiplicative interaction. The reference group is $RR_{no,no} = 1$ (in other words all our comparisons are to the risk among those who have neither A nor B). Note that $RR_{yes,yes} = RR_{yes,no} \times RR_{no,yes} = 2.0 \times 1.5 = 3.0$. ($RR_{yes,no}$ is the relative risk of B in the absence of A, and $RR_{no,yes}$ is the relative risk of A in the absence of B).

Relative Risk (RR)

	A no	A yes
B no	$RR_{no,no} = 1$	$RR_{no,yes} = 1.5$
B yes	$RR_{yes,no} = 2.0$	$RR_{yes,yes} = 3.0$

If our observed $RR_{yes,yes}$ is significantly different from 3.0 we can reject the null hypothesis of no interaction and conclude that there is a multiplicative interaction.

To test this statistically, we would include a product term in our regression model, (multiplying the value of variable B by the value of variable A for each person to get a new variable which is the product of A and B) and then calculate the following quantity:

$$\frac{\hat{\beta}(coefficient\ of\ the\ product\ term\ in\ the\ logistic\ regression)}{standard\ error\ of\ \hat{\beta}}$$

This quantity squared is approximately distributed as Chi-Square with 1 degree of freedom.

What we are really testing is whether the coefficient $\hat{\beta}$ is significantly different from 0. If it is, then this is equivalent to concluding that the $RR_{yes,yes}$ is significantly different from our expected value of 3.0

Suppose we wanted to see what kind of incidence figures might give rise to the table above. Remember, incidence is absolute risk, while relative risk is the absolute risk in one group relative to the absolute

risk in the reference group. The two tables below both contain incidence figures that would give rise to the RR table above, so you can see
it is possible to have different incidence rates or risks which have the
same relative risk.

Risk or Incidence per 1,000

	A no	A yes
B no	$I_{no,no} = 20$	$I_{no,yes} = 30$
B yes	$I_{yes,no} = 40$	$I_{yes,yes} = 60$

	A no	A yes
B no	$I_{no,no} = 10$	$I_{no,yes} = 15$
B yes	$I_{yes,no} = 20$	$I_{yes,yes} = 30$

Additive risk is less commonly tested for, although some people
think it should be. It is calculated from a difference in absolute risks
(rather than from the ratio of absolute risks). Under the hypothesis of
no interaction in an additive model, we would expect the data in the tables below.

Incidence per 1,000

	A no	A yes
B no	$I_{no,no} = 20$	$I_{no,yes} = 30$
B yes	$I_{yes,no} = 40$	$I_{yes,yes} = 50$

Note: $Incidence_{yes,yes}$ = Base incidence + effect of A + effect of B or

$$I_{yes,yes} = I_{no,no} + (I_{no,yes} - I_{no,no}) + (I_{yes,no} - I_{no,no}) = 20 + 10 + 20 = 50$$

Risk Differences or Attributable Risk (AR)

Per 1,000

	A no	A yes
B no	$AR_{no,no} = 0$	$AR_{no,yes} = 10$
B yes	$AR_{yes,no} = 20$	$AR_{yes,yes} = 30$

The $AR_{yes,yes}$ = effect of A plus effect of B = 10 + 20 = 30
If the $AR_{yes,yes}$ is sufficiently different from the expected value of 30, then we may conclude there is an interaction on the additive scale.

The relative risk table that corresponds to the incidence table for the example given above of the additive model is:

Relative Risk (RR)

	A no	A yes
B no	$RR_{no,no} = 1$	$RR_{no,yes} = 1.5$
B yes	$RR_{yes,no} = 2.0$	$RR_{yes,yes} = 2.5$

Thus, the expected value of $RR_{yes,yes}$ under the null hypothesis of no additive interaction is: $RR_{yes,yes} = RR_{yes,no} + RR_{no,yes} - 1$. If $RR_{yes,yes}$ is significantly different from the above expectation, we would be able to reject the null hypothesis of additive risk.

Interactions depend on the scale—i.e. whether we are talking about relative risks (multiplicative) or attributable risks (additive). It is wise to consult a statistician for appropriate interpretations. Excellent, more technical sources are *Epidemiology: Beyond the Basics* by Moyses Szklo and F. Javier Nieto and *Modern Epidemiology* by Rothman and Greenland.

Summary:

Additive model interaction effect:
See if observed value of $AR_{yes,yes}$ differs from expected value:

$$\mathrm{AR}_{\mathrm{yes,yes}} = \mathrm{AR}_{\mathrm{yes,no}} + \mathrm{AR}_{\mathrm{no,yes}}$$

or, $\mathrm{RR}_{\mathrm{yes,yes}} = \mathrm{RR}_{\mathrm{yes,no}} + \mathrm{RR}_{\mathrm{no,yes}} - 1$

Multiplicative model interaction effect:

See if observed value of $\mathrm{RR}_{\mathrm{yes,yes}}$ differs from expected value:

$$\mathrm{RR}_{\mathrm{yes,yes}} = \mathrm{RR}_{\mathrm{yes,no}} \times \mathrm{RR}_{\mathrm{no,yes}}$$

Chapter 5
MOSTLY ABOUT SCREENING

I had rather take my chance that some traitors will escape detection than spread abroad a spirit of general suspicion and distrust, which accepts rumor and gossip in place of undismayed and unintimidated inquiry.

Judge Learned Hand
October 1952

5.1 Sensitivity, Specificity, and Related Concepts

The issue in the use of screening or diagnostic tests is to strike the proper trade-off between the desire to detect the disease in people who really have it and the desire to avoid thinking you have detected it in people who really don't have it.

An important way to view diagnostic and screening tests is through sensitivity analysis. The definitions of relevant terms and symbols are as follows:

T+ means positive test, T– means negative test, D+ means having disease, D– means not having disease.

		True Condition Presence of Disease		
		Yes +	No –	
Diagnostic Test	+	a	b	$a + b$ = all testing positive
	–	c	d	$c + d$ = all testing negative
		$a + c$ = all diseased	$b + d$ = all nondiseased	$a + b + c + d$ = total population

129

SENSITIVITY: the proportion of diseased persons the test classifies as positive,

$$= \frac{a}{a + c} = P(T+ \mid D+); \quad \text{(probability of positive test, given disease)}$$

SPECIFICITY: the proportion of nondiseased persons the test classifies as negative,

$$= \frac{d}{b + d} = P(T- \mid D -); \quad \text{(probability of negative test, given no disease)}$$

FALSE-POSITIVE *RATE*: the proportion of nondiseased persons the test classifies (incorrectly) as positive,

$$= \frac{b}{b + d} = P(T+ \mid D -); \quad \text{(probability of positive test, given no disease)}$$

FALSE-NEGATIVE *RATE*: the proportion of diseased people the test classifies (incorrectly) as negative,

$$= \frac{c}{a + c} = P(T- \mid D+); \quad \text{(probability of negative test given disease)}$$

PREDICTIVE *VALUE OF A POSITIVE TEST*: the proportion of positive tests that identify diseased persons,

$$= \frac{a}{a + b} = P(D+ \mid T+); \quad \text{(probability of disease given positive test)}$$

PREDICTIVE *VALUE OF A NEGATIVE TEST*: the proportion of negative tests that correctly identify nondiseased people,

$$= \frac{d}{c + d} = P(D- \mid T -); \quad \text{(probability of no disease given negative test)}$$

ACCURACY *OF THE TEST*: the proportion of all tests that are correct classifications,

$$= \frac{a + d}{a + b + c + d}$$

LIKELIHOOD *RATIO OF POSITIVE TEST*: the ratio of probability of a positive test, given the disease, to the probability of a positive test, given no disease,

$$= \frac{P(T+\,|\,D+)}{P(T+\,|\,D-)} = \quad \text{positive test, given disease versus positive test, given no disease}$$

$$= \frac{\text{sensitivity}}{\text{false positive rate}} = \frac{\text{sensitivity}}{1 - \text{specificity}}$$

LIKELIHOOD *RATIO OF A NEGATIVE TEST*:

$$= \frac{P(T-\,|\,D+)}{P(T-\,|\,D-)} = \quad \text{negative test, given disease versus negative test, given no disease}$$

$$= \frac{1 - \text{sensitivity}}{\text{specificity}}$$

Note also the following relationships:

(1) Specificity + the false-positive rate = 1;

$$\frac{d}{b + d} + \frac{b}{b + d} = 1$$

therefore, if the specificity of a test is increased the false-positive rate is decreased.

(2) Sensitivity + false-negative rate = 1;

$$\frac{a}{a + c} + \frac{c}{a + c} = 1$$

therefore, if the sensitivity of a test is increased the false-negative rate will be decreased.

PRETEST *PROBABILITY OF DISEASE:* The pretest probability of a disease is its prevalence. Knowing nothing about an individual and in the absence of a diagnostic test, the best guess of the probability that the patient has the disease is the prevalence of the disease.

POSTTEST *PROBABILITY OF DISEASE:* After having the results of the test, the posttest probability of disease if the test is normal is $c/(c+d)$, and if it is abnormal the posttest probability is $a/(a+b)$. The last is the same as the *PREDICTIVE VALUE OF A POSITIVE TEST.*

A good diagnostic test is one that improves your guess about the patient's disease status over the guess you would make based on just the general prevalence of the disease. Of primary interest to a clinician, however, is the *predictive value of a positive test (PV+)*, which is the proportion of people who have a positive test who really have the disease, $a/(a+b)$, and the *predictive value of a negative test (PV-)*, which is the proportion of people with a negative test who really don't have the disease, $d/(c+d)$.

Sensitivity and specificity are characteristics of the test itself, but the predictive values are very much influenced by how common the disease is. For example, for a test with 95% sensitivity and 95% specificity used to diagnose a disease that occurs only in 1% of people (or 100 out of 10,000), we would have the following:

		Disease		
		Yes +	No −	
Test	+	95	495	590
	−	5	9,405	9,410
		100	9,900	10,000

The PV+ is 95/590 = .16; that means that only 16% of all people with positive test results really have the disease; 84% do not have the disease even though the test is positive. The PV- however, is 99.9%, meaning that if a patient has a negative test result, you can be almost completely certain that he really doesn't have the disease. The practical value of a diagnostic test is dependent on a combination of sensitivity, specificity, and disease prevalence, all of which determine the predictive values of test results.

If the prevalence of the disease is high, the predictive value of a positive test will also be high, but a good test should have a high predictive value even though the prevalence of the disease is low. Let us take a look at the relationship between disease prevalence and sensitivity, specificity and predictive value of a test, shown in Figure 5.1.

Let us, for instance, consider a test that has a sensitivity of .95 and a specificity of .99. That means that this test will correctly label as diseased 95% of individuals with the disease and will correctly label as nondiseased 99% of individuals without the disease. Let us consider a disease whose prevalence is 10%, that is, 10% of the population have this disease, and let us now look and see what the predictive value of a positive test is. We note that it is approximately .90, which means that 90% of individuals with a positive test will have the disease. We can see that the predictive value of a positive test, drops to approximately .70 for a test that has a sensitivity of .95 and a specificity of .95, and we can see that it further drops to approximately .40 for a test that has a sensitivity of .95 and a specificity of .85. In other words, only 40% of individuals with a positive test would truly have the disease for a test that has that particular sensitivity and specificity.

One thing you can note immediately is that *for disease of low prevalence, the predictive value of a positive test goes down rather sharply*. The other thing that you can notice almost immediately is that large difference in sensitivity makes a small difference in the predictive value of a positive test and that a small difference in specificity makes a big difference in the predictive value of a positive test. This means that the characteristic of a screening test described by specificity is more important in determining the predictive value of a positive test than is sensitivity.

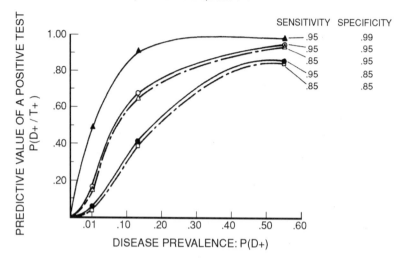

RELATIONSHIP BETWEEN DISEASE PREVALENCE,
SENSITIVITY, SPECIFICITY, AND PV

Figure 5.1

Figure 5.2 shows us a situation of a test that's virtually perfect. A test that has a sensitivity of .99 and a specificity of .99 is such that at most prevalence levels of disease the probability of disease, given a normal or negative test result, is very low. That would be a very good test, and the closer we can get to that kind of situation the better the diagnostic test is. The diagonal line in the center represents a test with a sensitivity of .50 and a specificity of .50 and that, of course, is a completely useless test because you can note that at the prevalence of the disease of .4 the probability of the disease given a positive test is also .4, which is the same as the probability of the disease without doing any test, and this pertains at each prevalence level. Therefore, such a test is completely useless whereas a test with sensitivity and specificity of .99 is excellent and anything in between represents different usefulness for tests. This then, is an analytic way to look at diagnostic test procedures.

A particularly relevant example of the implications of prevalence on

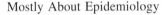

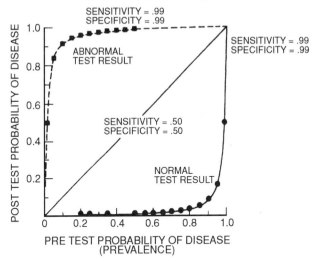

Figure 5.2

predictive value is the case of screening for the presence of infection with the AIDS virus. Since this disease is generally fatal, incurable at present, has a stigma attached to it, and entails high costs, one would not like to use a screening strategy that falsely labels people as positive for HIV, the AIDS virus.

Let us imagine that we have a test for this virus that has a sensitivity of 100% and a specificity of 99.995%, clearly a very good test. Suppose we apply it routinely to all female blood donors, in whom the prevalence is estimated to be very low, .01%. In comparison, suppose we also apply it to homosexual men in San Fran-cisco in whom the prevalence is estimated to be 50%.[22] For every 100,000 such people screened, we would have values as shown in the table on the following page.

Although in both groups all those who really had the disease would be identified, among female blood donors one third of all people who tested positive would really not have the disease; among male homo-

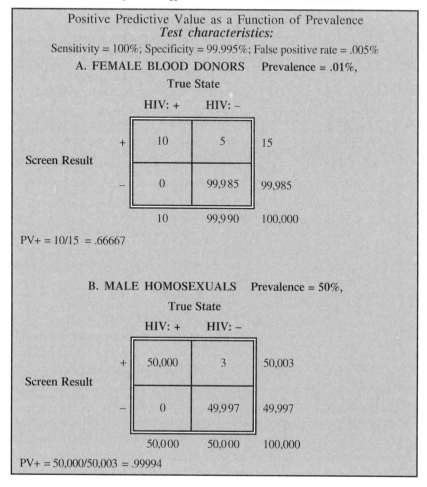

Positive Predictive Value as a Function of Prevalence
Test characteristics:
Sensitivity = 100%; Specificity = 99.995%; False positive rate = .005%

A. FEMALE BLOOD DONORS   Prevalence = .01%,

True State

	HIV: +	HIV: −	
Screen Result +	10	5	15
−	0	99,985	99,985
	10	99,990	100,000

PV+ = 10/15 = .66667

B. MALE HOMOSEXUALS   Prevalence = 50%,

True State

	HIV: +	HIV: −	
Screen Result +	50,000	3	50,003
−	0	49,997	49,997
	50,000	50,000	100,000

PV+ = 50,000/50,003 = .99994

sexuals only 6 out of 100,000 people with a positive test would be falsely labeled.

5.2 Cutoff Point and Its Effects on Sensitivity and Specificity

We have been discussing sensitivity and specificity as characteristic of a diagnostic test; however, they can be modified by the choice of the *cut-*

off point between normal and abnormal. For example, we may want to diagnose patients as hypertensive or normotensive by their diastolic blood pressure. Let us say that anyone with a diastolic pressure of 90 mm Hg or more will be classified as "hypertensive." Since blood pressure is a continuous and variable characteristic, on any one measurement a usually nonhypertensive individual may have a diastolic blood pressure of 90 mm Hg or more, and similarly a truly hypertensive individual may have a single measure less than 90 mm Hg. With a cutoff point of 90 mm Hg, we will classify some nonhypertensive individuals as hypertensive and these will be false positives. We will also label some hypertensive individuals as normotensive and these will be false negatives. If we had a more stringent cutoff point, say, 105 mm Hg, we would classify fewer nonhypertensives as hypertensive since fewer normotensive individuals would have such a high reading (and have fewer false positives).

However, we would have more false negatives (i.e., more of our truly hypertensive people might register as having diastolic blood pressure less than 105 mm Hg on any single occasion). These concepts are illustrated in Figure 5.3.

There are two population distributions, the diseased and nondiseased, and they overlap on the measure of interest, whether it is blood pressure, blood glucose, or other laboratory values. There are very few screening tests that have no overlap between normal and diseased individuals.

One objective in deciding on a cutoff point is to strike the proper balance between false positives and false negatives. As you can see in Figure 5.3, when the cutoff point is at A, all values to the right of A are called positive (patient is considered to have the disease). In fact, however, the patient with a value at the right of cutoff A could come from the population of nondiseased people, since a proportion of people who are perfectly normal may still have values higher than those above A, as seen in the normal curve. The area to the right of A under the no-disease curve represents the false positive.

If an individual has a test value to the left of cutoff A, he may be a true negative or he may be a false negative because a proportion of in-

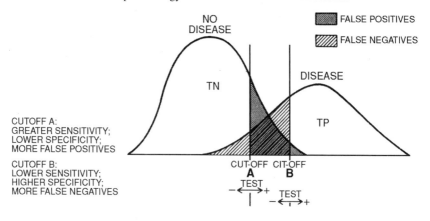

Figure 5.3

dividuals with the disease can still have values lower than cutoff A. The area under the "disease" curve to the left of cutoff A represents the proportion of false negatives.

If we move the cutoff point from A to B, we see that we decrease the area to the right of the cutoff, thereby decreasing the number of false positives, but increasing the number of false negatives. Correspondingly, with cutoff A, we have a greater probability of identifying the truly diseased correctly, that is, pick up more true positive, thereby giving the test with cutoff A greater sensitivity. With cutoff B, we are less likely to pick up the true positive (lower sensitivity) but more likely to correctly identify the true negatives (higher specificity).

Thus, by shifting the cutoff point beyond what we call a test positive, we can change the sensitivity and specificity characteristics of the test. The choice of cutoff, unless there is some special physiological reason, may be based on consideration of the relative consequences of having too many false positives or too many false negatives. In a screening test for cancer, for example, it would be desirable to have a test of high sensitivity (and few false negatives), since failure to detect this condition early is often fatal. In a mass screening test for a less serious condition or for one where early detection is not critical, it may be more desirable to have a high specificity in order not to overburden the health care delivery system with too many false positives. Cost consideration may also enter into the choice of cutoff point.

The relationship between sensitivity (the ability to correctly identify the diseased individuals) and the false-positive fractions is shown in Figure 5.4.

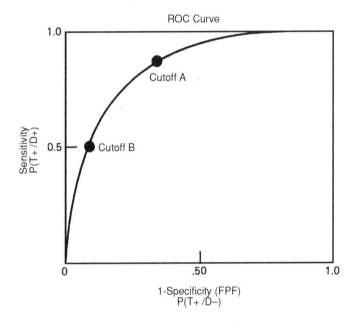

Figure 5.4

This is called the receiver operating characteristic (ROC) curve of the test. Often we can select the cutoff point between normal and abnormal depending on the trade-off we are willing to make between sensitivity and the proportion of false positives.

We can see that with cutoff A, while we can detect a greater percentage of truly diseased individuals, we will also have a greater proportion of false-positive results, while with cutoff B we will have fewer false positives but will be less likely to detect the truly diseased. Screening tests should have corresponding ROC curves drawn.

Chapter 6
MOSTLY ABOUT CLINICAL TRIALS

It is no easy task to pitch one's way from truth to truth through besetting errors.

Peter Marc Latham
1789–1875

"I wouldn't have seen it if I didn't believe it!"

Attributed to Yogi Berra

Unfortunately sometimes scientists see what they believe instead of believing what they see. Randomized, controlled clinical trials are intended to avoid that, and other kinds, of bias.

A *randomized clinical trial* is a prospective experiment to compare one or more interventions against a control group in order to determine the effectiveness of the interventions. A clinical trial may compare the value of a drug versus a placebo. A placebo is an inert substance that looks like the drug being tested. It may compare a new therapy with a currently standard therapy, surgical with medical intervention, two methods of teaching reading, two methods of psychotherapy. The principles apply to any situation in which the issue of who is exposed to which condition is under the control of the experimenter and the method of assignment is through randomization.

6.1 Features of Randomized Clinical Trials

(1) There is a group of patients who are designated study patients. All criteria must be set forth and met before a potential candidate can be considered eligible for the study. Any exclusions must be specified.

(2) Any reasons for excluding a potential patient from participating in the trial must be specified prior to starting the study. Otherwise, unintentional bias may enter. For example, supposing you are comparing coronary bypass surgery with the use of a new

drug for the treatment of coronary artery disease. Suppose a patient comes along who is eligible for the study and gets assigned
to the surgical treatment. Suppose you now discover the patient
has kidney disease. You decide to exclude him from the study because you think he may not survive the surgery with damaged
kidneys. If you end up systematically excluding all the sicker patients from the surgical treatment, you may bias the results in favor of the healthier patients, who have a better chance of survival
in any case. In this example, kidney disease should be an *exclusion criterion applied to the patients before they are assigned to
any treatment group.*

(3) Once a patient is eligible, he or she is randomly assigned to the
experimental or control group. Random assignment is not "hap-
hazard" assignment but rather it means that each person has an
equal chance of being an experimental or control patient. It is
usually accomplished by the use of a table of random numbers,
described later, or by computer-generated random numbers.

(4) Clinical trials may be double-blind, in which neither the treating
physician nor the patient knows whether the patient is getting the
experimental treatment or the placebo; they may be single-blind,
in which the treating physician knows which group the patient is
in but the patient does not know. A double-blind study contains
the least bias but sometimes is not possible to do for ethical or
practical reasons. For example, the doctor may need to know the
group to which the patient belongs so that medication may be
adjusted for the welfare of the patient. There are also trials in
which both patients and physicians know the treatment group, as
in trials comparing radical mastectomy versus lumpectomy for
treatment of breast cancer. When mortality is the outcome the
possible bias introduced is minimal, provided that exclusion criteria were specified and applied before eligibility was finally determined and that the randomization of eligible participants to
treatment groups was appropriately done.

(5) While clinical trials often compare a drug or treatment with placebo, they may also compare two treatments with each other, or a
treatment and "usual care". Trials that compare an intervention

with "usual care", obviously cannot be blinded, for example comparing a weight-loss nutritional intervention with "usual" diet; however, the assessment of effect (measurement of weight, or blood pressure, or some hypothesized effect of weight loss) should be done in a blinded fashion, with the assessor not knowing which group the participant has been assigned to.

(6) It is essential that the control group be as similar to the treatment group as possible so that differences in outcome can be attributed to differences in treatment and not to different characteristics of the two groups. Randomization helps to achieve this comparability.

(7) We are concerned here with Phase III trials. New drugs have to undergo Phase I and II trials, which determine toxicity, and safety and efficacy, respectively. These studies are done on small numbers of volunteers. Phase III trials are large clinical trials, large enough to provide an answer to the question of whether the drug tested is better than placebo or than a comparison drug.

6.2 Purposes of Randomization

The basic principle in designing clinical trials or any scientific investigation is to *avoid systematic bias*. When it is not known which variables may affect the outcome of an experiment, the best way to avoid systematic bias is to assign individuals into groups randomly. *Randomization* is intended to insure an approximately equal distribution of variables among the various groups of individuals being studied. For instance, if you are studying the effect of an antidiabetic drug and you know that cardiac risk factors affect mortality among diabetics, you would not want all the patients in the control group to have heart disease, since that would clearly bias the results. By assigning patients randomly to the drug and the control group, you can expect that the distribution of patients with cardiac problems will be comparable in the two groups. Since there are many variables that are unknown but may have a bearing on the results, randomization is insurance against unknown and unintentional bias. Of course, when dealing with variables known to be relevant, one can take these into account by *stratifying and then randomizing within the strata*. For instance, age is a variable relevant to diabetes outcome. To stratify by age, you might select four age groups for your study: 35–44, 45–54, 55–64, 65 plus. Each group is considered a stratum. When a patient enters into the clinical trial his age stratum is first determined and then he is randomly assigned to either experimental or control groups. Sex is another variable that is often handled by stratification.

Another purpose of randomization has to do with the fact that the statistical techniques used to compare results among the groups of patients under study are valid under certain assumptions arising out of randomization. The mathematical reasons for this can be found in the more advanced texts listed in the Suggested Readings.

It should be remembered that sometimes randomization fails to result in comparable groups due to chance. This can present a major problem in the interpretation of results, since differences in outcome may reflect differences in the composition of the groups on baseline characteristics rather than the effect of intervention. Statistical methods are available to adjust for baseline characteristics that are known to be related to outcome. Some of these methods are logistic regression, Cox proportional hazards models, and multiple regression analyses.

6.3 How to Perform Randomized Assignment

Random assignment into an experimental group or a control group means that each eligible individual has an equal chance of being in each of the two groups. This is often accomplished by the use of random number tables. For example, an excerpt from such a table is shown below:

| 48461 | 70436 | 04282 |
| 76537 | 59584 | 69173 |

Its use might be as follows. All even-numbered persons are assigned to the treatment group and all odd-numbered persons are assigned to the control groups. The first person to enter the study is given the first number in the list, the next person gets the next number and so on. Thus, the first person is given number 48461, which is an odd number and assigns the patient to the control group. The next person is given 76537; this is also an odd number so he too belongs to the control group. The next three people to enter the study all have even numbers and they are in the experimental group. In the long run, there will be an equal number of patients in each of the two groups.

6.4   Two-Tailed Tests Versus One-Tailed Test

A clinical trial is designed to test a particular hypothesis. One often sees this phrase in research articles: "Significant at the .05 level, two-tailed test." Recall that in a previous section we discussed the concept of the "null hypothesis," which states that there is no difference between two groups on a measure of interest. We said that in order to test this hypothesis we would gather data so that we could decide whether we should reject the hypothesis of no difference in favor of some alternate hypothesis. *A two-tailed test versus a one-tailed test refers to the alternate hypothesis posed.* For example, suppose you are interested in comparing the mean cholesterol-level of a group treated with a cholesterol-lowering drug to the mean of a control group given a placebo. You would collect the appropriate data from a well-designed study and you would set up the null hypothesis as

H_o: mean cholesterol in treated group = mean cholesterol in control group.

You may choose as the alternate hypothesis

H_A: mean cholesterol in treated group is *greater than* the mean in controls.

Under this circumstance, you would reject the null hypothesis in favor of the alternate hypothesis if the observed mean for women was sufficiently *greater* than the observed mean for men, to lead you to the conclusion that such a great difference in that direction is not likely to have occurred by chance alone. This, then, would be a one-tailed test of the null hypothesis.

If, however, your alternate hypothesis was that the mean cholesterol level for females is *different* from the mean cholesterol level for males, then you would reject the null hypothesis in favor of the alternate *either* if the mean for women was *sufficiently greater* than the mean for men *or* if the mean for women was *sufficiently lower* than the mean for men. The direction of the difference is not specified. In medical research it is more common to use a two-tailed test of significance since we often do not know in which direction a difference may turn out to be, even though we may think we know before we start the experiment. In any case, it is important to report whether we are using a one-tailed or a two-tailed test.

6.5   Clinical Trial as "Gold Standard"

Sometimes observational study evidence can lead to misleading conclusions about the efficacy or safety of a treatment, only to be overturned by clinical trials evidence, with enormous public health implications. The Women's Health Initiative (WHI) clinical trial of hormone therapy is a dramatic example of that.[21] Estrogen was approved by the FDA for relief of post-menopausal symptoms in 1942, aggressively marketed in the mid 1960's, and after 1980, generally combined with progestin for women with a uterus because it was found that progestin offset the risks of estrogen for uterine cancer. In the meantime many large prospective follow-up studies almost uniformly showed that estrogen reduced heart diseases by 30-50%. In the 1993 WHI mounted a large clinical trial to really answer the question of long-term risks and benefits of hormone therapy. One part was the study of estrogen alone for women had had a hysterectomy, and thus didn't need progestin to protect their uterus, and another part was of estrogen plus progestin (E+P) for women with an intact uterus.

The E+P trial was a randomized, double blind, placebo-controlled clinical trial meant to run for an average of 8.5 years. It included 16,608 women ages 50-79; such a large sample size was deemed necessary to obtain adequate power. The trial was stopped in 2002, three years before its planned completion, because the Data and Safety Monitoring Board or DSMB, (as described in Chapter 9) found estrogen plus progestin caused an excess of breast cancer, and surprisingly, there was a significant and entirely unexpected excess of heart attacks in the E+P group compared to placebo! Final results, reported in subsequent papers, showed that the adverse effects (a 24% increase in invasive breast cancer, 31% increase in strokes, 29% increase in coronary heart disease and more than a two-fold increase in pulmonary embolism and in dementia) offset the benefits, (a 37% decrease in colorectal cancer and 34% decrease in hip fractures), so that taken together, the number of excess harmful events per year was substantial. Since there were 6 million women taking this preparation in the United States alone, and millions more globally, these results have important implications for women other than those in the trial itself.

Why such different results from a clinical trial than from observational longitudinal studies? The most likely explanation is selection bias. Women who were taking hormones and then followed to observe their rates of heart disease, were in virtually all the observational studies, healthier, thinner, more active, more educated, less overweight, than their non-hormone taking counterparts, and their healthier lifestyle and better baseline health status, rather than the hormones per se, was what accounted for their lower rates of heart disease.

The question now is answered using the "gold standard," the clinical trial: estrogen plus progestin does not protect against heart disease, and in fact increases the risk. As noted before, the impact of this research is great since so many millions of women were using the preparation tested.

6.6   Regression Toward the Mean

When you select from a population those individuals who have high blood pressure and then at a later time measure their blood pressure again, the average of the second measurements will tend to be lower than the average of the first measurements and will be closer to the mean of the original population from which these individuals were drawn. If between the first and second measurements you have instituted some treatment, you may incorrectly attribute the decline of average blood pressure in the group to the effects of treatment, whereas part of that decline may be due to the phenomenon called *regression toward the mean*. (That is one reason why a placebo control group is most important for comparison of effects of treatment above and beyond that caused by regression to the mean.) Regression to the mean occurs when you select out a group because individuals have values that fall above some criterion level, as in screening. It is due to variability of measurement error. Consider blood pressure.

The observed value of blood pressure is the person's true value plus some unknown amount of error. The assumption is that people's measured blood pressure is normally distributed around the mean of their true but unknown value of blood pressure. Suppose we will only take people into our study if their blood pressure is 160 or more. Now suppose someone's true systolic blood pressure is 150, but we measure it 160. We select that person for our study group just because his measured value is high. However, the next time we measure his blood pressure, he is likely to be closer to his true value of 150 than the first time. (If he had been close to his true value of 150 the first time, we would never have selected him for our study to begin with, since he would have been below our cutoff point. So he must have had a large error at that first measurement.) Since these errors are normally distributed around his true mean of 150, the next time we are more likely to get a lower error and thus a lower measured blood pressure than the 160 that caused us to select him.

Suppose now that we select an entire subgroup of people who have high values. The averages of the second measurements of these selected people will tend to be lower than the average of their first measurements, and closer to the average of the entire group from which we selected them. The point is that people who have the highest values the first time do not always have the highest values the second time because the correlation between the first and second measurement is not perfect. Similarly, if we select out a group of people because of low values on some characteristic, the average of the second measurements on these people will be higher than the average of their first measurements, and again closer to the mean of the whole group.

Another explanation of this phenomenon may be illustrated by the following example of tossing a die. Imagine that you toss a die 360 times. Whenever the die lands on a five or a six, you will toss the die again. We are interested in three different averages: (1) the mean of the first 360 tosses, (2) the mean of the tosses that will result in our tossing again, and (3) the mean of the second tosses. Our results are shown in the table on the next page.

Although on the first toss the mean of the 360 times is 3.5, we only pick the two highest numbers and *their* mean is 5.5. These 120 times when the die landed on 5 or 6 will cause us to toss again, but on the second toss the result can freely vary between 1 and 6. Therefore, the mean of the second toss must be lower than the mean of the group we selected from on the first toss specifically because it had the high values.

First Toss		Second Toss	
Result	# of Times Result Is Obtained	Result	# of Times Result Is Obtained
1	60		
2	60		
3	60		
4	60		
5	60	1	20
6	60	2	20
		3	20
		4	20
		5	20
Mean of 360 tosses = 3.5		6	20
Mean of the 120 tosses that landed 5 or 6 = 5.5		Mean of the 2nd toss = 3.5	

6.7 Intention-to-Treat Analysis

Data from clinical trials in general should be analyzed by comparing the groups as they were originally randomized, and not by comparing to the placebo control group only those in the drug group who actually did take the drug. The people assigned to the active drug group should be included with that group for analysis even if they never took the drug. This may sound strange, since how can one assess the efficacy of a drug if the patient isn't taking it? But the very reason people may not comply with the drug regimen may have to do with adverse effects of the drug, so that if we select out only those who do comply we have a different group from the one randomized and we may have a biased picture of the drug effects.

Another aspect is that there may be some quality of compliers in general that affects outcome. A famous example of misleading conclusions that could arise from not doing an intention-to-treat analysis comes from the Coronary Drug Project.[26] This randomized, double-blind study compared the drug clofibrate to placebo for reducing cholesterol. The outcome variable, which was five-year mortality, was very similar in both groups, 18% in the drug group and 19% in the placebo group. It turned out, however, that only about two thirds of the patients who were supposed to take clofibrate actually were compliant and did take their medication. These people had a 15% mortality rate, significantly lower than the 19% mortality in the placebo group. However, further analysis showed that among those assigned to the placebo group, one third didn't take their placebo pills either. The two thirds of the placebo group who were compliant had a mortality of 15%, just like the ones who complied with the clofibrate drug! The noncompliant people in both the drug and placebo groups had a higher mortality (25% for clofibrate and 28% for placebo). It may be desirable in some circumstances to look at the effect of a drug in those who actually take it. In that case the comparison of drug compliers should be to placebo compliers rather than to the placebo group as a whole.

The inclusion of non-compliers in the analysis dilutes the effects, so every effort should be made to minimize noncompliance. In some trials a judged capacity for compliance is an enrollment criterion and an evaluation is made of every patient as part of determining his or her eligibility as to whether this patient is likely to adhere to the regimen. Those not likely to do so are excluded prior to randomization. However, if the question at hand is how acceptable is the treatment to the patient, in addition to its efficacy, then the basis for inclusion may be the general population who might benefit from the drug, including the non-compliers.

In the Women's Health Initiative, the primary analysis was intention-to-treat. However a secondary analysis adjusted for compliance (more commonly referred to as adherence). In this analysis the event history of the participant was censored six months after she either stopped taking the study pills or was taking less than 80% of the study pills. In the placebo group the event history was censored six months after the participant started taking hormones (some participants in the placebo group stopped taking study pills but were prescribed hormones by their physicians and started taking them on their own). Thus this secondary analysis basically compared the two groups "as treated" rather than as assigned to a particular treatment. In the intention-to-treat analysis the hazard ratio for coronary heart disease was 1.24 while in the "adherence adjusted" analysis it was 1.50. Thus the findings from the intention-to-treat analysis were confirmed and strengthened in the adherence-adjusted analyses.

6.8   How Large Should the Clinical Trial Be?

A clinical trial should be large enough, that is, have big enough sample size, to have a high likelihood of detecting a *true* difference between the two groups. If you do a small trial and find no significant difference, you have gained no new information; you may not have found a difference simply because you didn't have enough people in the study. You cannot make the statement that there is no difference between the treatments. If you have a large trial and find no significant difference, then you are able to say with more certainty that the treatments are really not different.

Suppose you do find a significant difference in a small trial with p < .05 (level of significance). This means that the result you obtained is likely to have arisen purely by chance less than 5 times in 100 (if there really were no difference). Is it to be trusted as much as the same p value from a large trial? There are several schools of thought about this.

The p value is an index of the strength of the evidence with regard to rejecting a null hypothesis. Some people think that a p value is a p value and carries the same weight regardless of whether it comes from a large or small study. Others believe that if you get a significant result in a small trial, it means that the effect (or the difference between two population means) must be large enough so that you were able to detect it even with your small samples, and therefore, it is a meaningful difference. It is true that if the sample size is large enough, we may find statistical significance if the real difference between means is very, very small and practically irrelevant. Therefore, finding a significant difference in a small trial does mean that the effect was relatively large.

Still others say that in practice, however, you can have less confidence that the treatments really do differ for a given p value in a small trial than if you had obtained the same p value in testing these two treatments in a large trial.[27] This apparent paradox may arise in situations where there are many more small trials being carried out worldwide studying the same issue than there are large trials—such as in cancer therapy. Some of those trials, by chance alone, will turn out to have significant results that may be misleading.

Suppose that there are 1,000 small trials of anticancer drug therapy. By chance alone, 5% of these will be significant even if the therapies have no effect, or 50 significant results. Since these are by chance alone, it means we are incorrect to declare anticancer drug effects in these trials (we have committed type I errors). Suppose, further, that there are only 100 large trials studying this same issue. Of these, 5%, or five such studies, will declare a difference to exist, incorrectly. So if we combine all the trials that show significant differences *incorrectly*, we have 55 such significant but misleading p values. Of these, 50 or 91% come from small trials and 5 out of the 55 incorrect ones (or 9%) come from the large trials. The following points are important:

(1) There is a distinction between *statistical significance* and *clinical significance*. A result may not have arisen by chance, that is, it may reflect a true difference, but be so small as to render it of no practical importance.

(2) It is best to report the actual probability of obtaining the result by chance alone under the null hypothesis, that is, the *actual p value*, rather than just saying it is significant or not. The *p* value for what we commonly call "significance" is arbitrary. By custom, it has been taken to be a *p* value of .05 or less. But the .05 cutoff point is not sacred. The reader should decide what strength he or she will put in the evidence provided by the study, and the reader must have the information to make that decision. The information must include the design of the study, sample selection, the sample sizes, the standard deviations, and the actual *p* values.

In summary:

(1) Finding *no significant difference* from a small trial tells us nothing.
(2) Finding *no significant difference* in a large trial is a real finding and tells us the treatments are likely to be equivalent.
(3) Finding a *significant difference* in a small trial may or may not be replicable.
(4) Finding a *significant difference* in a large trial is to be trusted as revealing a true difference.

6.9   What Is Involved in Sample Size Calculation?

a. Effect size

Let us say that 15% of victims of a certain type of heart attack die if they are given drug A and 16% die if they are given drug B. Does this 1% difference mean drug A is better? Most people would say this is too small a difference, even if it doesn't arise by chance, to have any clinical importance. Suppose the difference between the two drugs is 5%.

Would we now say drug A is better? That would depend on how large a difference we thought was important. The size of the difference we want to detect is called the *effect size*.

To calculate sample size you need to know the minimum size of the difference between two treatments that you would be willing to *miss* detecting. Suppose for example that in your control group 30% of the patients without the treatment recover. It is your belief that with treatment in the experimental group 40% will recover. You think this difference in recovery rate is clinically important and you want to be sure that you can detect a difference at least as large as the difference between 30% and 40%. This means that if the treatment group recovery rate were 35% you would be willing to miss finding that small an effect. However, if the treatment rate was 40% or more, you would want to be pretty sure to find it. How sure would you want to be? The issue of "how sure" has to do with the "power" of the statistical test.

b. Power

Statistical power means the *probability* of finding a real effect (of the size that you think is clinically important). The relationships among power, significance level, and type I and type II error are summarized below:

Significance level = probability of a type I error = probability of finding an effect when there really isn't one. This is also known as alpha or α.

Probability of type II error = probability of failing to find an effect when there really is one. This is also known as beta or β.

Power = 1 − probability of type II error = probability of finding an effect when there really is one. This is also known as *1 − beta*.

c. Sample size

To calculate sample size, you have to specify your choice of effect size, significance level, and desired power. If you choose a significance level of .05 and a power of .80, then your type II error probability is 1 – power or .20. This means that you consider a type I error to be four times more serious than a type II error (.20/.05 = 4) or that you are four times as afraid of finding something that isn't there as of failing to find something that is. When you calculate sample size there is always a trade-off. If you want to decrease the probability of making a type I error, then for a given sample size and effect size you will increase the probability of making a type II error. You can keep both types of error low by increasing your sample size. The top part of the table on the next page shows the sample sizes necessary to compare two groups with a test between two proportions under different assumptions.

The second row of the table shows that if you want to be able to detect a difference in response rate from 30% in the control group to 50% or more in the treatment group with a probability (power) of .80, you would need 73 people in each of the two groups. If, however, you want to be fairly sure that you find a difference as small as the one between 30% and 40%, then you must have 280 people in each group.

If you want to be more sure of finding the difference, say 90% sure instead of 80% sure, then you will need 388 people in each group (rather than the 280 for .80 power). If you want to have a more stringent significance level of .01, you will need 118 people in each group (compared with the 73 needed for the .05 significance level) to be able to detect the difference between 30% and 50%; you will need 455 people (compared with 280 for the .05 level) to detect a difference from 30% to 40% response rate.

The bottom part of the table on the next page shows the impact on sample size of a one-tailed test of significance versus a two-tailed test. Recall that a two-tailed test postulates that the response rate in the treatment group can be *either larger or smaller* than the response rate in the control group, whereas a one-tailed test specifies the direction of the hypothesized difference. *A two-tailed test requires a larger sample size*, but that is the one most commonly used.

Sample Size Examples				
Significance Level (1-tailed)	Assume:	Effect Size	Power	Sample Size
	Control Group Response Rate =	Detect Increase in Treatment Group at Least to:	With Probability of:	n Needed in Each Group
.05	30%	40%	.80	280
	30%	50%	.80	73
	30%	40%	.90	388
	30%	50%	.90	101
.01	30%	40%	.80	455
	30%	50%	.80	118
	30%	40%	.90	590
	30%	40%	.90	153

Sample Size Examples				
Significance Level = .05	Assume:	Effect Size	Power	Sample Size
	Control Group Response Rate =	Detect Increase in Treatment Group at Least to:	With Probability of:	n Needed in Each Group
1-Tailed	30%	40%	.80	280
2-Tailed	30%	40%	.80	356
1-Tailed	30%	50%	.80	73
2-Tailed	30%	50%	.80	92

d. Some additional considerations

For a fixed sample size and a given effect size or difference you want to detect, maximum power occurs when the event rate is about 50%. So to maximize power it may sometimes be wise to select a group for study that is likely to have the events of interest. For example, if you want to study the effects of a beta-blocker drug on preventing heart attacks, you could get "more power for the money" by studying persons who have already had one heart attack rather than healthy persons, since the former are more likely to have another event (heart attack). Of course you might then be looking at a different question, the effect of beta-blockers on survivors of heart attack, (which would be a secondary prevention trail) rather than the effect of beta-blockers in preventing the first heart attack.(a primary prevention trial). Sometimes a primary prevention trial gives a different answer than a secondary prevention trial. You may be able to intervene to prevent disease among people not yet suffering from the disease, but your intervention may have little effect on someone who has already developed the disease. Clearly judgment is required.

6.10 How to Calculate Sample Size for the Difference Between Two Proportions

You need to specify what you think the proportion of events is likely to be in each of the two groups being compared. An event may be a response, a death, a recovery—but it must be a dichotomous variable. Your specification of the event rates in the two groups reflects the size of the difference you would like to be able to detect.

Specify:

p_1 = rate in group 1; $q_1 = 1 - p_1$; alpha = significance level

p_2 = rate in group 2; $q_2 = 1 - p_2$; power

$$n = \frac{(p_1 q_1) + (p_2 q_2)}{(p_2 - p_1)^2} \times f(alpha, power)$$

The values of f (alpha, power) for a two-tailed test can be obtained from the table below.

Values of f (alpha, power)

		.95	.90	.80	.50
Alpha	.10	10.8	8.6	6.2	2.7
Significance					
Level .01	.05	13.0	10.5	7.9	3.8
	.01	17.8	14.9	11.7	6.6

Note: n is roughly inversely proportional to $(p_2 - p_1)^2$.

Example. Suppose you want to find the sample size to detect a difference from 30% to 40% between two groups, with a power of .80 and a significance level of .05. Then,

$$p_1 = .30; \quad q_1 = .70; \quad \text{alpha} = .05$$
$$p_2 = .40; \quad q_2 = .60; \quad \text{power} = .80$$

$$f(\text{alpha, power}) = 7.9 \text{ from the table}$$

$$n = \frac{(.30)(.70) + (.40)(.60)}{(.40 - .30)^2} \times 7.9 = 356$$

You would need 356 people in each group to be 80% sure you can detect a difference from 30% to 40% at the .05 level.

6.11 How to Calculate Sample Size for Testing the Difference Between Two Means

The formula to calculate sample size for a test of the difference between two means, assuming there is to be an equal number in each group, is

$$n = \frac{k \times 2\sigma^2}{(MD)^2} = number\ in\ each\ group$$

where σ^2 is the error variance, MD is the minimum difference one wishes to detect, and k depends on the significance level and power desired. Selected values of k are shown on the next page. For example, to detect a difference in mean I.Q. of 5 points between two groups of people, where the variance $= 16^2 = 256$, at a significance level of .05 and with power of .80, we would need

$$n = \frac{7.849 \times 2(256)}{(5)^2} = 161 \text{ people}$$

Significance Level	Power	k
.05	.99	18.372
	.95	12.995
	.90	10.507
	.80	7.849
.01	.99	24.031
	.95	17.814
	.90	14.879
	.80	11.679

in each group, or a total sample size of 322. This means we are 80% likely to detect a difference as large or larger than 5 points. For a 10-point difference, we would need 54 people in each group.

A common set of parameters for such sample size calculations are $\alpha = .05$ and power $= .80$. However, when there are multiple comparisons, we have to set α at lower levels as described in Section 3.24 on the Bonferroni procedure. Then our sample size would need to be greater.

If we are hoping to show that two treatments are equivalent, we have to set the minimum difference we want to detect to be very small and the power to be very, very high, resulting in very large sample sizes.

To calculate values of k that are not tabulated here, the reader is referred to the book *Methods in Observational Epidemiology* by Kelsey, Thompson, and Evans for an excellent explanation. There are computer programs avaiable to calculate power for many different situations. An excellent on is: NCSS (National Council for Social Studies statistical software) which can be obtained by going to the website: www.ncss.com.

Chapter 7
MOSTLY ABOUT QUALITY OF LIFE

The life which is unexamined is not worth living.

Plato, Dialogues
428–348 B.C.

I love long life better than figs.

Shakespeare
(Anthony and Cleopatra)

The two quotes above illustrate how differently people view the quality of their lives and how difficult it is to pin down this concept.

A welcome development in health care research is the increasing attention being paid to quality of life issues in epidemiological studies and when evaluating competing therapies. A key aspect is the measurement of the effects of symptoms of illness, as well as of the *treatment* of these symptoms, on well-being, which is a subjective and relative state. Therefore, it is quite appropriate that measurement of improvement or deterioration in quality of life be based on the *patient's perception and self-report.* A person who has had severe and disabling angina may perceive improved well-being as a result of treatment if he can walk without pain, whereas a young ski enthusiast may experience marked deterioration if he is unable to ski. For that reason, in studies on this issue the individual often serves as his or her own control, and the measures used are *change scores* in some quality of life dimensions from before to after treatment. However, it remains important to have an appropriate control group to compare the changes, because people show changes in these dimensions over time that may be unrelated to the particular treatment being evaluated.

The principles and techniques described in this book apply to research in any health-related field. However, there are certain analytic methods that are particularly appropriate to investigations concerning

161

psychological or emotional states. The primary principle is that if it is to be scientific research, it must adhere to scientific standards, which means that first of all, *the variables of interest must be quantified.* Fortunately, almost any concept related to the health fields can be quantified if one is ingenious enough.

7.1 Scale Construction

The scales used to measure quality of life dimensions reflect the degree of distress with particular symptoms or psychological states as well as degree of satisfaction and general well-being. There are many such scales available, which have been well constructed and tested on different populations. Sometimes, however, investigators find it necessary to construct their own scales to fit particular circumstances.

There are three characteristics of such scales that are important: *reliability, validity*, and *responsiveness.*

7.2 Reliability

Reliability is the ability to measure something the same way twice. It rests on the assumption that a person's score on a scale or a test is composed of his true (but unknown) score plus some component that is subject to variation because of error (by which we mean random variability).

Reliability of a scale is related to its *repeatability,* or how close the responses are on two administrations of the scale. To measure how close they are we can calculate the correlation coefficient between the two administrations of the scale to the same subjects. But often we can't give the same scale to our patients twice under exactly the same circumstances, since in reality a patient responding twice to the same questions would respond differently either because something has intervened between the two occasions or because he remembered the previous responses or just because there is inherent variability in how one feels. The next best thing would be to give two equivalent scales to the same group, but that has its problems as well. How do we know the two scales are really equivalent?

Fortunately, there are various measures of what we call "internal consistency" that give us the reliability of a scale or test. The most common one is called Cronbach's alpha. There are many software packages for personal computers that readily calculate Cronbach's alpha, including SPSS, SAS, STATA, and many others. Thus, it is not necessary to calculate it yourself, but the following explanation indicates what it really means and how to interpret it.

$$\alpha = \left[\frac{k}{k-1} \right] \times \left[\frac{variance\ of\ total\ scale\ -\ sum\ of\ variances\ of\ individual\ items}{variance\ of\ total\ scale} \right]$$

Variance is the standard deviation squared. Section 3.4 shows how to calculate it. (When we talk about variance here, we actually mean the population variance, but what we really use are estimates of the population variance that we get from the particular sample of people on whom we develop the test or scale, since obviously we can't measure the entire population.)

This formula is really a measure of how homogeneous the scale items are, that is, to what extent they measure the same thing. If you have a scale that is composed of several different subscales, each measuring different things, then the Cronbach's alpha should be used for each of the subscales separately rather than the whole scale. Cronbach's alpha gives the lower bound for reliability. If it is high for the whole scale, then you know the scale is reliable (repeatable, highly correlated with the "true," but unknown, scores). If you get a low alpha for the whole scale, then either it is unreliable or it measures several different things.

Reliability can also be looked upon as a measure of correlation, and in fact it does reflect the average correlation among items of a scale, taking into account the total number of items. Another way to get reliability is from the Spearman–Brown formula, which is

$$\frac{k(average\ correlation\ among\ all\ items)}{1 + (k-1)\ average\ correlation\ among\ all\ items} = \frac{k\ (r_{average})}{1 + (k-1)(r_{average})}$$

As this formula indicates, a longer test or scale is generally more reliable if the additional items measure the same thing. On the other hand, shorter scales are more acceptable to patients. An alpha above .80 is considered very good, and sometimes subscales are acceptable with alpha over .50, particularly when there are a large number of subjects (over 300), but it should be considered in the context of the other psychometric qualities of the scale.

There are other measures of reliability as well. Psychometrics is a specialized and complex field and there are many excellent books on the subject, for example, *Health Measurement Scales* by Streiner and Norman.

7.3 Validity

Validity refers to the degree to which the test measures what it is supposed to measure. An ideal situation would be one in which there is some external criterion against which to judge the measuring instrument, a "gold standard." For example, if it could be shown that anxiety as measured on one scale correlates better with some objectively definable and agreed upon outcome than anxiety measured on a second scale, one could say the first scale is more valid. (This is called "criterion validity.")

Unfortunately, in quality of life issues there are generally no external criteria. A person may feel he or she is miserable, but be functioning at a high level. The very idea of quality of life is conceptually subjective. Whose quality of life is it anyway?

Therefore, we often must rely on content validity, which is a blend of common sense and technical psychometric properties. If we want to know if someone feels depressed we might ask, "Do you feel sad a great deal?" rather than, "Do you feel athletic?" However, even that is not so simple, since what someone who is not an expert on depression might consider overtly irrelevant, like sleep disturbances, is one of the most powerful signs of depression.

Of course, if there is an external criterion against which to validate a scale, it should be used. But even content validity may be made more objective, for instance by forming a group of experts to make judgments on the content validity of items. To test the agreement between judges, the kappa coefficient may be used, as described in Section 3.3.

7.4 Responsiveness

Responsiveness of a scale is a measure of how well it can detect changes in response to some intervention. Responsiveness, or sensitivity of a scale, can be assessed in several different ways and there is no consensus as to which is the best. Some related concepts are described below, which pertain to the situation when you are looking at change from pre- and posttreatment measures.

(1) The use of *change scores* (pre-post) is appropriate when the variability between patients is greater than the variability within patients. In general, change scores can safely be used if

$$\frac{\sigma^2_{between\ patients}}{\sigma^2_{between} + \sigma^2_{error}} \geq 0.5$$

$\sigma^2_{between\ patients}$ and σ^2_{error} can be obtained from an analysis of variance of scores of a group of patients who have replicated measures, so that you can estimate the variance due to error.

(2) A *coefficient of sensitivity* to change due to a treatment is

$$\frac{\sigma^2_{change}}{\sigma^2_{change} + \sigma^2_{error}}$$

To get the σ^2_{error}, one needs to do an analysis of variance of repeated measures on the same subjects. Computer programs are available. Detailed explanations of this appear in more advanced texts.

(3) *Effect size* is simply the change in the scale from before to after treatment, divided by the standard deviation at baseline. The standard deviation is an index of the general variability in scores among the group of people in the study. One can measure the magnitude of the average change in scores after some treatment by determining what percentage of the "background variation" that change represents. *Effect size* =

$$\frac{mean\ change\ score}{standard\ deviation\ of\ baseline\ (or\ pretreatment)\ scores}$$

(4) A *measure of responsiveness* proposed by Guyatt et al.[25] is

$$\frac{mean\ change\ score}{standard\ deviation\ of\ change\ scores\ for\ ``stable\ subjects"}$$

"Stable subjects" are hard to define, but what this suggests is that a control group that doesn't get the intervention or gets placebo may be used. Then one can use the standard deviation of the change scores in the control group as the denominator in the term above.

The variability of the change scores in the control group (or in a group of stable subjects) can be looked at as the "background variability" of *changes* and the measuring instrument is responsive to the degree it can detect changes above and beyond this background variability.

(5) When evaluating change due to treatment, one should *always* have a control group (i.e., a no-intervention or placebo group) for comparison, since change can occur in control patients as well, and the question of interest is whether the pre- to posttreatment change in

the treatment group exceeds the "background" change in the control group. If you use effect size as a measure, then you should compare effect size in the treatment group with effect size in the control group.

A numerical example of these concepts is provided in Appendix G.

7.5 Some Potential Pitfalls

a. Multiplicity of variables

Quality of life research often deals with a vast quantity of variables. Let us say an investigator is examining the effects of a drug to treat hypertension and comparing it with placebo. The investigator may have several hundred items to assess various physical and psychological symptoms and side effects. If one were to compare the two groups by t-test on each of the items, at the $p = .05$ level of significance, one would expect that roughly 5% of these tests would produce a significant result by chance alone. The exact probability is difficult to determine, since some of these comparisons would be correlated by virtue of the fact that the same patients are responding to all of them, that is, the responses are not independent. But in any case, if the investigators pick out just the significant items and conclude that there are effects of the drug, they may be committing type I errors, that is, rejecting the null hypothesis incorrectly.

That is why it is important to use scales that measure particular constructs, or to group items in a clinically meaningful way. For example, one might wish to measure depression, anxiety, hostility, wellbeing (each of which consists of multiple items). On the other hand, certain drugs may be related to very specific symptoms, such as nightmares, and this might need to be assessed by a single item that asks about the frequency of nightmares.

The point is that quality of life research should generally be driven by some specific hypotheses. Otherwise it becomes a "fishing expedition" that just fishes around for anything significant it can find. It should be noted that "fishing expeditions" may be useful to generate hypotheses that then need to be tested in a different study.

b. Generalization

Another important issue is the extrapolation of results to populations other than the one from which the study sample was drawn. Quality of life effects may be different in men than in women, in younger than in older people, and may differ by ethnic and cultural groups. One should be careful in making generalizations.

c. Need for rigorous standards of research

Some people consider quality of life measures "soft." What they generally mean is that they think such measures are subjective, variable, and perhaps meaningless. That is nonsense, and to the extent it is true in some studies it reflects the inadequacies of the researcher, not of the subject matter. These measures *should be subjective* from the patient's perspective, since they reflect the patient's subjective perception of well-being or distress. *It is the researcher who should not be subjective,* and who need not be if he follows the principles of research. The variability in quality of life measures is no greater than in many physiologic measures, and, in any case, is part of the essence of some quality of life constructs. As for meaning, that is a philosophical issue, not a scientific one. From the scientific viewpoint, the "meaning" should be defined operationally. Quality of life research should adhere to the principles of all good research and the general approach is the same as for any scientific investigation:

(1) formulate a testable hypothesis;
(2) quantify the dependent variable (or variables);
(3) select a study design that can answer the question you've posed;
(4) quantify the independent variables;
(5) control for potential confounders (through study design and/or data analysis);

(6)   plan for a sample size that will give you enough power to detect an effect size of interest;

(7)   try to ensure that you minimize systematic bias;

(8)   collect the data, paying much attention to quality control;

(9)   analyze the data using appropriate statistical techniques; and

(10) make inferences consistent with the strengths and limitations of the study.

Chapter 8
MOSTLY ABOUT GENETIC EPIDEMIOLOGY

Let us then suppose the mind to be, as we say, white paper (tabula rasa), void of all characters without any ideas; how comes it to be furnished? Whence comes it by that vast store, which the busy and boundless fancy of man has painted on it with an almost endless variety?....To this I answer, in one word, From experience: in that all our knowledge is founded....

John Locke
An Essay Concerning Human Understanding (1689)

8.1 A New Scientific Era

We are a long way from believing that the mind is a "tabula rasa," a blank slate. We know now that much is in fact innate, i.e. under genetic influence. The purpose of this chapter is to help those who wish to read the rapidly expanding literature in genetic epidemiology. Thus, it is an overview of the basic designs and statistics used in this area; it is not comprehensive, nor is it highly technical. Appendix H provides basic descriptions and definitions of the intracellular process under control of genes, for those unfamiliar with the area.

The focus of epidemiological research has evolved as parallel progress has been made in other fields of medicine and basic science. In the era when infectious diseases were rampant, epidemiology was concerned with identifying the sources of the infection and methods of transmission, largely through fieldwork. As the infectious agents were discovered, as sanitation and health status improved, chronic diseases, such as heart disease and cancer, became the leading causes of death and disability in the developed world, and came to be the foremost targets of epidemiological research. (Now that new infectious diseases are once again emerging, this part of epidemiology is again gaining prominence.)

The objective of chronic disease epidemiology was to identify risk factors for these diseases. This part of the story has been a great public

171

health success. We now know, because of epidemiological studies, what the major modifiable risk factors are for cardiovascular disease: hypertension, high cholesterol and low HDL, smoking, overweight, and inactivity. Our challenge now is to find ways to make the lifestyle changes in the population, which will further lower the rates of cardiovascular disease. We also know many of the exposures related to cancer, but not as comprehensively as for heart disease.

At this scientifically historic time, as science is fully entering into the era of *genomics and proteomics*, epidemiology is entering into a new phase of research activity: *molecular epidemiology*. This is the search for blood or tissue biomarkers and genetic polymorphisms (variants) that are associated with, or pre-dispose to disease. Why is this different from any other risk factor investigated in epidemiology? In many ways it isn't, especially with regard to the blood biomarkers, but in genetic epidemiology there are study designs and statistical analysis methods that are quite different. A really new aspect of molecular and genetic epidemiology is the true collaboration of basic scientists, clinicians and epidemiologists. For too long the disciplines have gone their separate research ways and scientists read mostly the scientific journals in their own field. But molecular epidemiology cannot fruitfully proceed without the interface of laboratory scientists and population researchers.

8.2   Overview of Genetic Epidemiology

Genetic epidemiology seeks to identify genes related to disease and to assess the impact of genetic factors on population health and disease. Here is an overview of the strategy often used to study genetic determinants of disease. First we may want to determine if the disease *runs in families*. If it is not familial, it is not likely to be heritable; if it is familial, it may or may not be due to genetic factors (environments run in families also). Next, we want to see if genes contribute to the familial transmission. One method for determining this is by studying twins (described in Section 8.3). If we determine the *disease is heritable*, we would want to *localize and identify* the genes involved.

As a first step, we may want to find out where the genes that contribute to the disorder are located. This can be done by genome scan linkage studies of individuals affected with the disease and their families (described in Section 8.4). *Linkage studies* may identify regions on the chromosome that are likely to harbor the disease genes. Once we've identified one or more such regions, we may look to see what genes are known to reside in those regions. We can then test these genes using *association studies in unrelated individuals* (described in Section 8.6) to determine whether any variants (also called *alleles*) of these genes are associated with the disease.

So, there are a variety of designs and statistical tests that can be used to define the genetic basis of a disease, including: 1) *twin studies* to determine if the disease has a heritable component; 2) *linkage studies* to identify and locate regions of chromosomes containing genes involved in the disease; and 3) *association studies* to determine whether specific genetic variants are associated with the disease, to examine how they interact with the environment, and to determine how they affect population health. We will limit the discussion to some pretty simple models that will give the flavor of the topic. Readers interested in more depth are referred to the many more technical writings on the subject, but particularly to the excellent Special Report on genetics by Ellsworth and Manolio.[28,30,31]

8.3 Twin Studies

To establish a genetic influence on disease, we may look to see if it runs in families. But something that is familial is not necessarily hereditary. For example: do obese parents have obese children because of genetics or because of nutrition and activity levels that are transmitted from the parents to the children? What we want to know is whether and to what extent the phenotype (what we observe in the person, for example, obesity) is affected by genetic factors.

One way to assess the influence of heredity is from studies of twins. Identical twins (monozygotic—coming from the same fertilized egg) share 100% of their genes, while fraternal twins (dizygotic—coming from two fertilized eggs) share only 50%, just as non-twin siblings do. One way to estimate the strength of genetic influences is by a *heritability index* h^2. This is commonly defined as twice the difference between the correlation for that trait among monozygotic twins minus the correlation in dizygotic twins or:

$$h^2 = 2(r_{mz} - r_{dz})$$

Consider blood pressure. If variation in the condition or trait under investigation were completely attributable to genetic variation, then each member of a monozygotic twin pair would be equally affected (each member would have the same blood pressure) and the correlation between monozygotic twins would be 1.0; the correlation in dizygotic twins however, would be .50.

In this case, h^2 would be $2(1 - 0.5) = 1.0$ or 100%. If the condition is completely not heritable in the population, then $r_{mz} = r_{dz}$ and $h^2 = 0$. Since diseases and traits are generally partially heritable, h^2 lies somewhere between 0 and 1.0.

If we are talking about continuous variables, we can think of heritability in terms of correlation coefficients. If we are talking about categorical variables, we may speak of concordance rates, where

$$h^2 = \frac{\%\ monozygotic\ twins\ concordant\ for\ the\ disease - \%\ of\ dizygotic\ twins\ concordant}{1 - \%\ of\ dizygotic\ twins\ concordant}$$

Some reported approximate estimates[32,33,34,35] of heritability are: 1.0 for Huntington's disease (this is because Huntington's disease is a single gene disorder, inherited in a dominant mode of transmission, and fully penetrant—this term is defined toward the end of this chapter); .60 for alcoholism; .40 for the personality trait of conscientiousness; .35 for colorectal cancer;.26 for multiple sclerosis; and .80 for schizophrenia.

It is important to remember that heritability doesn't measure how much of an individual's disease is attributable to genetics; rather it tells us what proportion of the population's variability in the phenotype is the result of variation in the genes in the population. So it is a measure applicable to a population, not to an individual. If you have people living in exactly the same environment, then any variation you encounter in the phenotype must be due to genetic factors, since there is no environmental variation. In such a case, if all environmental factors are constant for the population, heritability would be 100%. So there are some limitations to this measure, but it does give us an idea to what extent genetic variation contributes to phenotypic variation in a population. However, heritability tells us nothing about what genes are responsible for that variation, what chromosomes they are on, where on the chromosome they are located, or what polymorphisms are involved.

8.4   Linkage and Association Studies

If we know a disease is heritable, we can now turn to the task of actually identifying the genes that are involved. Most disorders that are studied by epidemiologists (e.g. cardiovascular diseases, psychiatric disorders, common forms of cancer) are considered "complex" disorders. That is, unlike single-gene or Mendelian disorders, such as cystic fibrosis or Huntington's disease, these diseases are thought to result from the contribution of several or many genes interacting with environmental risk factors. That can make identifying the effect of an individual gene quite a difficult task. The effect of a particular allele within that gene may be quite small. It is a bit like looking for the proverbial needle in the haystack. Nevertheless, genes contributing to diseases are being discovered and there are certain strategies that are employed in the search.

Where in the genome do we look for the genes that confer susceptibility to the disease? One way to answer this question is to use *genetic linkage analysis*.

   a) Linkage analysis relies on the phenomena of crossing over and recombination that occur during the process of meiosis when the sex cells (sperm and egg) are formed. Each person has two copies of each of the 23 chromosomes that make up the genome: one copy is inherited from the mother and one from the father. During the formation of sperm and egg cells, these 23 chromosome pairs line up and exchange segments of genetic material in a process known as *crossing over*. This recombination occurs at one or more places along the chromosome. *The closer two loci are on a chromosome, the less likely a recombination event will occur between them and so the more likely they will be inherited together.* Loci that tend to be co-inherited are said to be genetically linked. We can use this fact to estimate the distance between two genetic loci or markers (a genetic marker is a piece of DNA whose chromosomal location is known). The physical distance between two markers is inversely related to how frequently they are co-inherited across generations in a family.

b) The *distance between two loci is sometimes measured in centi-Morgans*. A centiMorgan (cM) is a unit of distance along a chromosome, but not in the ordinary sense of physical distance. It is really a probability measure which is a reflection of the physical distance; it reflects the probability of two markers or loci being separated (or segregated) by crossing over during meiosis. If the two markers are very close together, they won't separate (we say they are "linked"); if they are far apart, they are likelier to cross over and *the genetic material gets recombined during meiosis*. Then this recombined DNA gets transmitted to the offspring. Two loci are one centimorgan apart if the probability that they are separated by crossing over is only 1% (once in a hundred meioses). It has been estimated that there are about 1 million base pairs in a 1cM span. Loci that are far apart, say 50 cM, will be inherited independently of each other, as they would be if they were on different chromosomes. *The purpose of linkage studies is to localize the disease-susceptibility gene to be within some region on the chromosome.*

c) So we might begin our search for a disease gene by collecting families affected by the disease and performing a linkage analysis using markers spaced at intervals (say 10 cM apart) across the entire genome. This is known as *a whole genome scan*. If we find a marker that appears to be co-transmitted with the disease, we would have evidence that the marker is genetically linked to a gene for the disease. In other words, there is likely to be a gene for the disease in the same region as the linked marker (there's a pony in there somewhere.*)

d) Having found a chromosomal region linked to the disease, we might try to narrow the region down by genotyping and testing additional markers within that region (say at 1 cM intervals). However, even this relatively small region may contain many genes.

e) Our next step might be to screen the genes that are known to reside in this region. We would be particularly interested in genes that have a plausible connection to the disease of interest (these would be good *"candidate genes"*). For example, if we are studying diabetes, genes that make proteins involved in glucose metabolism would be important "candidate genes."

f) Now we can see if any particular alleles (variants) of the genes in that chromosomal region are associated with the disease. This can be done by:

1. *Association studies using case-control methods in unrelated people*, examining whether an allele is more common in cases than controls (described in Section 8.6);

2. *Association studies in families* to see whether an allele is being transmitted more commonly to cases than expected by chance (described in Section 8.7).

---

* This refers to the story of a pair of twins. One was the optimist and one the pessimist. The psychologist tried to cure them of their respective distortions and so he put the pessimist twin into a room full of toys. The child cried inconsolably, explaining that he was crying because he would never be able to select a toy to play with from among this whole big heap. The psychologist put the optimist twin in a room full of horse manure, and when he came back half an hour later, the child was laughing and clapping his hands with joy. "How come you're so happy here", asked the psychologist, and the optimistic twin replied, "I figure with all this horse manure, there's got be a pony in there somewhere."

So linkage analysis tells us that a particular marker location is near a disease susceptibility gene; association analysis tells us that a particular allele of a gene or marker is more commonly inherited by individuals with the disease.

8.5   LOD Score: Linkage Statistic

The classic statistic used to evaluate the strength of the evidence in favor of linkage of a genetic marker and disease susceptibility gene is the LOD score (the $\log_{10}$ of the odds in favor of linkage). It will be described in principle only, to help in interpretation of epidemiological articles dealing with genetics. The actual calculations are complex and require special program packages.

The principle underlying the LOD score is described in the previous section: if we have two loci—say, a marker and a disease gene—the closer they are on a chromosome, the lower the probability that they will be separated by a recombination event during meiosis and the more likely they will be co-inherited by offspring.

The probability of recombination, called the recombination fraction, is denoted by the symbol θ and depends on the distance between the gene and the marker. *If there is no recombination and the gene and marker are completely linked, then the recombination fraction is 0.* The maximum value of θ is .5 (if gene and marker were independently inherited, then the probability that the marker was transmitted but not the gene, equals the probability that gene was transmitted but not the marker, equals .50).

So if you want to know if there is linkage, we have to estimate how likely it is that θ is less than .5, given the data we have observed. We use the likelihood ratio for this, which as you recall from Chapter 2 is the ratio of the probability of observed symptoms, given disease divided by the probability of observed symptoms given no disease. In this case:

$$LR = \frac{Probability\ observed\ inheritance\ data,\ given\ linkage}{Probability\ observed\ inheritance\ data,\ given\ no\ linkage}$$

The null hypothesis here is no linkage (or recombination fraction θ = .5) and the alternate hypothesis is linkage (or θ <.5). If we reject the null, we "accept" the alternate hypothesis. The test statistics used to see if we have sufficient data to conclude linkage is the *LOD score which is the* $\log_{10}$ *(LR.).* For Mendelian (single gene) disorders, a lod score of 3 has traditionally been the threshold for declaring significant linkage, although for complex disorders higher thresholds (3.3 – 3.6) have been recommended. A LOD score of 3.0 indicates 10^3 odds in favor of linkage compared to no linkage, i.e. 1000:1 odds in favor of linkage.

For complex reasons beyond the scope of this book (but described in the references at the end), a LOD score can be translated into probability by multiplying it by the constant 4.6: LOD x 4.6 is distributed as chi-square with 1 degree of freedom. (The 4.6 is 2 times the natural log of 10.) Thus a LOD of 3.0 is equivalent to a chi-square of 3 x 4.6 =13.82, and corresponds to p = .0002. The inheritance data for linkage analyses can come from family pedigree studies, from sibships or other family groups.

LOD score linkage analysis is sometimes referred to as "parametric" linkage analysis because it requires that we specify certain parameters (e.g. disease and marker allele frequencies, recessive vs. dominant mode of inheritance, penetrance of the disease gene). When these parameters are known or can be approximated, parametric LOD score analysis is the most powerful method of linkage analysis. This may be true for Mendelian (single gene) disease, but for many complex disorders, these parameters are not known. "Nonparametric" linkage methods (known as the allele-sharing approach) are often used to study complex disorders because they do not require knowledge of the mode of inheritance or other genetic parameters. There are a number of statistics available, described in the more advanced texts.

8.6   Association Studies

Compared to linkage analysis, association studies are more closely akin to traditional epidemiological studies. They may be used to evaluate a candidate gene we are investigating when previous data, generally from linkage studies, have suggested that a particular gene is involved and maybe even a particular variant in the gene. In these studies, investigators are interested in finding whether there is any association between a particular allele of a polymorphic gene and the disease in question. (A polymorphism is a variation at a particular locus on the chromosome). For the purposes of this discussion, we will assume that the polymorphisms we are looking at are *SNPs (single nucleotide polymorphisms)* or variants in a single one of the bases A T C G at a particular locus (See Appendix H for details on SNPs).

So let us say at a particular SNP some people have the allele A and other people have the allele G. We want to know if people with the disease are more likely to have say, the A allele than the G allele. We can do case-control studies of association by taking cases who are affected with the disease and unrelated controls who are not. We then test statistically whether the proportion of affected individuals who have allele A is greater than the proportion of controls who have allele A.

If the unit of observation is the person, we can analyze the data by genotype (i.e. compare proportions of cases and controls with one allele from the mother and one from the father resulting in their genotype of AA, AG, vs GG,). We can use ordinary statistical tests of the differences between proportions, or multiple logistic regressions (see Section 4.16) to determine the odds ratio connected with the allele in question, and we can test for gene-environment interactions by including an interaction term of the presence of the allele and some environmental factor, such as smoking. Association studies can be more powerful than linkage analysis for detecting genes of modest effect, making them an attractive approach for studying complex disorders which are expected to involve multiple genes of relatively small individual effect.

A potential problem related to such studies is the choice of control groups. Ethnic differences in the prevalence of the allele, which may be unrelated to the presence of disease and which have arisen because the allele has been transmitted across generations within ethnic groups may confound our results. This phenomenon is referred to as *"population stratification."* Therefore, first of all, we would need to know about the background distribution of these alleles in different ethnic population subgroups.

For example, let us say the A variant is more common in Caucasians and the G variant is more common in African Americans. (Such differences between groups can happen because of different progenitors in the different groups.) If it happens that our disease group has more Caucasians and our no-disease control group has more African Americans, then we might find that allele A is more common among those with disease than in those without disease, but it might really just be a reflection of the fact that we had more Caucasians cases than controls and the Caucasians are more likely to be carriers of the A allele. So we would have a false-positive finding because of the ethnic composition of the two groups. But ethnicity is not so simply determined, because both Caucasians and African Americans have multiple ethnic origins and may have different progenitors, and different patterns of alleles normally occurring.

Thus, if we find an association between a particular allele and the disease we are studying, it may be for one of four reasons: (1) it could be a false positive due to chance (Type I error); (2) it could be a false positive due to confounding because of population stratification; 3) it may be that the allele is in "linkage disequilibrium" with the true disease allele, meaning that the allele we found more frequently in cases than in controls, is located physically close enough to the true disease allele that the two alleles tend to be inherited together and co-occur in affected individuals; 4) there really is a true causal association of the allele we studied and the disease. (As we said before, genetics—and life—are not simple.)

8.7   Transmission Disequilibrium Tests (TDT)

A statistical test of both linkage and association is the *transmission disequilibrium test*. This is a family-based association test which avoids confounding due to population stratification by examining the transmission of alleles *within* families. If a marker and a polymorphism in a disease-susceptibility gene are in *linkage disequilibrium*, it means that they are so tightly linked that specific alleles of the marker tend to be inherited togther with specific alleles of the gene. Let's again consider a SNP at which there are two possible alleles, A and G. A parent who is heterozygous at this SNP (i.e. has genotype AG) can transmit either an A or a G to the child. Under the null hypothesis of no linkage and no association, the probability of transmitting either of these alleles is 50%. If we observe transmission of the A allele significantly more often than chance expectation (i.e. more than 50% of the time) to the affected offspring, then we conclude that the A allele is associated with the disease.

The basic statistics are fairly straightforward. The unit of analysis is a *transmission of an allele*. Here is an example using trios. A trio is an affected offspring and both parents. There are two transmissions possible of an allele in a particular locus—one from the mother and one from the father. In the diagram below we construct a 2x2 table and count the number of transmitted alleles that belong in each cell of the table.

Trio 1: since the affected child has two AA alleles in the locus under consideration, she had to have received an A from each parent; thus we see that the father transmitted his A allele and not his G and the mother also transmitted her A allele and not her G, indicating that these two transmissions belong in cell b which describes the transmission of an A and not a G allele.

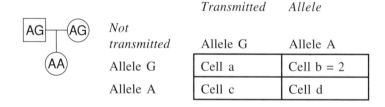

		Transmitted	Allele
	Not transmitted	Allele G	Allele A
	Allele G	Cell a	Cell b = 2
	Allele A	Cell c	Cell d

Trio 2: Here one transmission was of an A and not a G and the other was of a G and not an A. (We don't know which was from which parent, but we do know that 1 transmission is described by cell b and the other by cell c.)

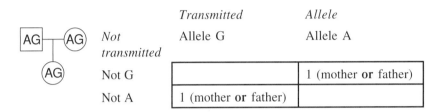

Not transmitted	Transmitted	Allele
	Allele G	Allele A
Not G		1 (mother **or** father)
Not A	1 (mother **or** father)	

Trio 3: Here, the A in the child had to have come from the father since that was all he had to transmit, so he transmitted an A allele and also did not transmit his other A allele; thus that transmission belongs in cell d. The mother transmitted her G allele and not her A allele and so that transmission belongs in cell c.

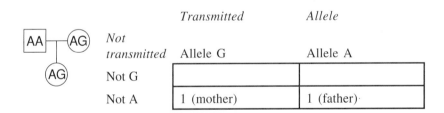

Not transmitted	Transmitted	Allele
	Allele G	Allele A
Not G		
Not A	1 (mother)	1 (father)

Imagine we have 120 such trios. We would now combine the data from the 240 transmissions among the 120 trios by adding the numbers in each cell into a summary table, as shown in the example below:

Not transmitted	Transmitted	Allele
	Allele G	Allele A
Not G	65	90
Not A	50	45

and calculate the test statistic TDT as the quantity:

$$\frac{(b-c)^2}{(b+c)} = \frac{(90-50)^2}{140} = 11.43$$

which is distributed as χ^2, with one degree of freedom, and since it is more than 3.84, we can reject the hypothesis of no linkage and conclude there is evidence of linkage and association. The TDT is really a McNemar's test (described in Section 3.2), analogous to matched case-control analysis, where controls are untransmitted alleles, rather than persons.

Note that any transmissions that land in the a or d cell are non-informative. The test statistic TDT only uses information from the b and c cells. Thus, only parents who are heterozygous (having an A and G) are informative. Since we start out with affected children identified by phenotype, we don't know whether the parents are heterozygous and must genotype all the parents in our collection of trios even though some will turn out not to be informative. Also, in our examples of trios, the affected children each had at least one A allele, but it possible for the child to be affected and have a GG genotype because there may be other genes that confer disease suscpetibility.

The sample size required for sufficient power to detect linkage and association through the TDT is dependent on many factors, including the marker and disease allele frequency, the recombination fraction or linkage disequilibrium between the marker and the disease allele and the effect size of the disease allele. A genetic statistician is best able to calculate the sample size and should be consulted before any such work is undertaken.[34]

The TDT has been extended to be applicable to cases where one or both parents' DNA is not available, to sibships and to other family groups, as well as to quantitative traits.

To make it more concrete, imagine a study of the genetics of schizophrenia. Let's say that previous studies have linked schizophrenia to a region on chromosome 22 and that there is a gene in this region that we consider a strong candidate gene because it is involved in the production of a neurotransmitter thought to be involved in the disease. We further know that there is a SNP in this gene that affects the amino acid sequence of the protein made by the gene. We want to test whether this SNP is associated with schizophrenia. We have a sample of 200 trios consisting of an offspring with schizophrenia and both parents. Each parent can transmit one of his or her two alleles at this locus to the affected offspring. For each trio, we can construct a 2x2 table and count the number of transmitted alleles that belong in each cell of the table, and then combine the tables to get a summary table that we use to calculate the TDT statistic as shown above. If that statistic is greater than 3.84 we can reject the hypothesis of no linkage at the .05 level significance.

8.8   Some Additional Concepts and Complexities of Genetic Studies

How do we select *candidate genes?* One strategy is to look at the biological pathway involved in the disease, consider a relevant protein in that pathway, and explore gene polymorphisms related to that protein. Another possibility is to look at genes within a region on the chromosome that is linked to the disease, even if we don't know what these genes actually do. Association studies look at candidate genes, but we don't have compelling candidates for many diseases. Technology is available to do *whole genome association scans, using association methods to test alleles by examining* polymorphisms throughout the whole genome. However, covering the genome in this way may require testing many thousands of SNPs, creating a problem of multiple hypothesis testing and a high probability of Type I error. (See Section 3.24.)

Haplotype analyses and *haplotype maps* are promising avenues to reduce the number of polymorphisms that have to be examined to find disease susceptibility genes. Haplotypes are sequences of alleles along a given chromosome. The concept of linkage disequilibrium tells us that alleles that are physically close together may be inherited together. These high linkage disequilibrium haplotypes, or "haplotype blocks," may be conserved through generations because they are inherited as a "block" and recent studies have shown that there is limited haplotype diversity in the human population. It turns out that there are several alleles within such haplotypes that characterize the entire haplotype and it may be possible to just look at those key alleles instead of at each single nucleotide polymorphism separately.

Among things that complicate genetic studies are the following:

It is often difficult to identify *a phenotype* specifically enough. Say we are looking for the gene for hypertension—how do we define a hypertensive when blood pressure is a continuous variable? Hypertension has been defined as being above a certain cut-point of blood pressure based on predictions of morbidity and mortality risks associated with different levels, but that does not necessarily correspond to some heritable characteristics.

Or take genes pre-disposing to heart disease. You have to define heart disease very specifically—does it include heart failure (which may be a different disease process); does it only include early heart attacks? In psychiatry, diagnoses are often based on having "x" number of symptoms out of a possible "n" symptoms, but what about people who have "x − 1" such symptoms? The diagnostic category may not be the heritable one. One approach to these problems of phenotype defintion is to identify underlying components or intermediate phenotypes that may be a more direct expression of gene effects than are the complex diagnoses used in medicine. Examples of such intermediate phenotypes include IgE levels in genetic studies of asthma, functional brain imaging in genetic studies of schizophrenia, and lipoprotein levels in studies of atherosclerotic heart disease.

Most diseases have complex modes of inheritance. There is not one gene that determines susceptibility, but several, each of which contributes a modest amount and which may interact with each other. The *penetrance* of the genes varies. This means that even if you have the disease allele you may not have the disease, either because other genes are also involved or because specific environmental factors are a necessary condition for the gene to operate. Very few genes are completely penetrant; Huntington's disease is one that is: people who have the Huntington's disease gene, will get the disease.

You may be looking at phenotypes which are really *phenocopies*. Phenocopies are phenotypes of the disease which are not caused by genetic factors but rather by environmental factors. Inadvertently including phenocopies in your analyses will act as noise and dilute any findings related to true genetic influences.

Polymorphisms in different genes may produce very similar phenotypes. This is known as genetic *heterogeneity*. Further complications arise from *epistasis* which occurs when the expression of the disease gene is affected by the action of another gene. So, as you can see, the field is very complex and the statistics to evaluate whether findings of linkage and association are more than chance are also complex.[36,37,38] New technology (such as whole genome association scans) and new analytic tools (such as "hapmaps") are being rapidly developed to solve such problems.

There has been a paradigm shift in science—from believing things are simpler than they seem to understanding they are more complex than they seem. For the last century the principle guiding scientific endeavor was Occam's razor—that the most parsimonious explanation for phenomena is the best. But as genomic and molecular discoveries accelerate, it becomes apparent that in the biological sphere simple explanations are not possible and the aim is to more accurately uncover and explain the inherent complexity (and marvel) of life.

Chapter 9
RESEARCH ETHICS AND STATISTICS

Morality, like art, means drawing a line someplace.
                                         Oscar Wilde (1854–1900)

9.1   What does statistics have to do with it?

At first glance it may seem that statistics and research ethics have
nothing to do with each other. Not so! Consider why so many people
volunteer for medical research studies. In many cases it is because
there is an expected benefit. For example in cancer clinical trials often
the investigational drug is a last hope and may not be available outside
of the trial. In many cardiovascular disease studies, participants appreciate the additional care and attention and are willing to try a new drug,
for example, for hypertension. And in fact, it has been shown that often, clinical trial participants live longer and do better than the general
population even if they are treated with a placebo. But what is perhaps
not sufficiently appreciated is that many, many people participate in
studies out of altruism to advance scientific knowledge. Scientific
knowledge is not advanced when a study is poorly designed, or carried
out without sufficient rigor, or not large enough to give an answer.
Proper statistics are a determinant of the ethics of a study.

A prime example is the Women's Health Initiative (WHI), described in Chapter 6. Postmenopausal women were asked to join a
study of hormone replacement therapy; the study would continue for
up to 12 years before the results were known and might not directly
benefit the women themselves, but they would answer the important
question of the effect of hormones on cancer, heart disease and osteo-
porosis. Many of the WHI participants took part for their daughters
and granddaughters, and they expressed pride and enthusiasm for an-
swering the questions for future generations. And indeed they did
achieve that goal—one part of WHI, the estrogen plus progestin trial
versus placebo, has already answered these important questions, with a

startling result: estrogen plus progestin increases risk of breast cancer and also increases heart attacks, stroke and blood clots, and dementia. So although the treatment does show benefit with regard to colorectal cancer and osteoporotic fractures, the overall risks outweigh the benefits. This trial has changed medical practice for generations to come.

Well that brings us back to statistics. This study was able to answer these questions because it had sufficient power to answer them. It required 16,608 women in that part of WHI to be able to detect these effects. Even if the result had been null (i.e. if it showed no difference between the treatment and placebo groups), we could have had faith in that result because the power was there to detect a true effect if there really was one. As it turned out, the results were clear-cut, though unexpected, in favor of placebo. So the point is that in order for a study to be "ethical" it must be designed and powered well enough so that it can answer the questions it poses. Otherwise, people who consent to participate in the expectation that they will contribute to knowledge may actually not be contributing because the study is poorly designed, powered or executed, and may be needlessly exposed to risk.

Note that there are certain study designs for which power considerations are less relevant. Examples are pilot studies, which by definition are intended to test the feasibility of a research protocol or to gather preliminary data to plan a full study, are exempt from the power issue. Power considerations may also not apply to certain drug toxicity studies (Phase I trials) or certain types of cancer trials, But certainly in prevention trials, as well as Phase III treatment trials, power is a major consideration in the ethics of research.

9.2   Protection of Human Research Subjects

Human subjects in medical research contribute greatly to improving the health of people. These volunteers must be protected from harm as much as possible. In the not-too-distant past, there were some egregious breaches of ethical principles in carrying out medical research. The world's most appalling examples are the medical experiments carried out in the Nazi concentration camps—by doctors! It defies any kind of understanding how educated, presumably "civilized" professionals could so have distorted their profession and their own humanity. But these atrocities did occur and demonstrate the horrors people are capable of perpetrating. When these atrocities became known after World War II, the Nuremberg trials of Nazi war criminals (including the doctors who preformed such research) also resulted in the Nuremberg Code for conduct of medical research which established the basic requirement of voluntary informed consent. Subsequently, the Declaration of Helsinki, in 1964, expanded and refined the research guidelines, and became a world standard, which undergoes periodic revisions.

The most infamous example of unethical research in the U.S. was probably the Tuskegee Institute study of syphilis, which took place in the south in the U.S. from 1932 to 1972. The researchers wanted to study the natural course of syphilis. In the 1940's antibiotics became available which could treat this disease, but were withheld from the participants, who were poor Black men, because an intervention to treat the disease would interfere with this observational study. In 1972 the public became aware of this experiment and in 1974 the National Commission for the Protection of Human Subjects of Biomedical and Behavioral Research was established. They developed a report known as *The Belmont Report*: Ethical Principles and Guidelines for the Protection of Human Subjects of Research. These guidelines are followed by all medical schools and other research institutions that conduct research involving human participants, and they are deemed to be universal principles, cutting across cultural lines. The guidelines are based on three basic principles: *respect for persons, beneficence, and justice.*

Respect for persons recognizes that people are autonomous beings and can make their own informed choices about participating in research, free of coercion. The informed consent process is predicated on this principle. Participants who are not able to make their own choices, such as comatose patients, or mentally incapacitated persons, or young children, must have special protections.

Beneficence, or the principle of non-malfeasance, means that the risks of the research must be kept to a minimum, the benefits maximized, and the researcher is responsible for protecting the participant.

Justice in this context refers to a fair distribution of the risks and benefits of research. One group of people should not be exposed to research risks for the benefit of another group of people. This can get to be a pretty complicated concept. While it may be easy to discern breaches in certain situations—to take the most extreme example, prisoners of the Nazis were subjected to freezing experiments to benefit soldiers who might have to fight under arctic conditions—it may be more subtle in many situations and these must be examined carefully, according to this principle.

9.3   Informed Consent

One of the most important elements in protection of human subjects is the principle of informed consent. The study subject must freely consent to be part of the study after being fully informed of the potential risks and benefits.

There are certain elements that must be in a written consent form. The purpose of the research must be stated; a 24-hour contact person must be listed; there must be a description of the study procedures: what is expected of the participant, the duration of the study, and how much of the participant's time it will take. The potential risks and discomforts, potential benefits, inconvenience to the participants, all must be clearly stated. There must be a statement that participation is voluntary and that the participant has the right to withdraw at any time and that this will not prejudice the care of the participant. If the research may result in need for further care or diagnostic procedures, the participant must be told to what extent he or she is responsible for further care and what the study will pay for. If there is any compensation to the participants, either for expenses incurred in participating or time spent, they must be informed of the amount. (The amount should not be excessive, as that may appear coercive.) A statement assuring confidentiality and how it will be maintained must be included.

Most important, the participant must understand what he or she is agreeing to and the consent form must be phrased in language that is understandable, and if appropriate, translated into the participant's native language. All this must be approved by the medical institution's IRB (or Institutional Review Board), which is generally a committee of experts and lay people who review and must approve all research protocols before the research is started, and who monitor adverse events as the research progresses. Different IRBs have different specific requirements that are usually posted on their web sites. Informed consent is an ongoing process—it is not just signing a form at the beginning of a study. The researcher has an obligation to keep the participant informed of relevant new research that may affect his or her decision to continue participating.

Back to the WHI—since it was believed at the time WHI was started that hormones would protect women from heart disease, the initial consent form stated this as a potential benefit. Potential risks stated in the consent form included an increase in breast cancer and blood clots. When WHI was in progress, the HERS (Heart and Estrogen Replacement Study) published results indicating that for women who already had heart disease (secondary prevention trial), hormone replacement provided no benefit. They observed more heart attacks in the early part of the study, with a possible late reduction, resulting in no overall difference between the treatment and placebo groups by the end of the study. This information was provided by a special mailing to all women participating in the WHI hormone program for primary prevention of heart disease. (Primary prevention means the study was carried out in generally healthy women). Subsequently, early data from the WHI itself indicated there was early harm with respect to heart disease. Again, the women were informed by a special mailing, telephone and personal discussion with clinic staff. Ultimately, the estrogen plus progestin trial was stopped after 5.2 years (instead of the originally planned average of 8.5 years) because the excess breast cancer risk crossed a pre-determined stopping boundary and a global index of overall effects suggested more harm than benefit, and all women in the trial were discontinued from their study pills.

9.4 Equipoise

That brings us to another concept: when is it ethical to begin a clinical trial of a new treatment? When there is equipoise. *Equipoise* means that there is about equal evidence that the treatment may provide benefit as there is that it will not provide benefit. If we are sure the treatment provides benefit, we should not deny it to people who would be getting placebo in the trial. Of course we may be wrong. There were critics of the Women's Health Initiative who said it was unethical to do such a trial because it was well known that hormones protect against heart disease and it would be unethical to deny these hormones to the women randomized to placebo! Of course we now know that was wrong—the placebo group did better. At the time WHI was started, the observational evidence pointed to benefit with regard to heart disease, but it had never been tested in a clinical trial, which is the "gold standard." Thus, there were many people who did not believe that the benefits of hormone replacement were already established by the observational studies, and it turns out they were right. The researcher, whose obligation it is to protect human research participants, must believe it is equally likely that the treatment is better or that the placebo or comparison treatment is better. The scientific community that judges the research proposal must believe, based on the "state-of-the-art", that there is a reasonable question to be answered.

9.5 Research Integrity

For research conclusions to be valid, data collection procedures must be rigorously and uniformly administered. No data may be altered without documentation. If there is a clerical error, the change and reason for it must be documented. Enrollment must be according to strict and pre-planned standards. Sometimes (fortunately, rarely) there is a great pressure to enroll subjects in a given time frame, or the researcher (in violation of the principle of equipoise) really believes the treatment can help his or her patients, and so "bends" the enrollment rules. This may invalidate the research and so is unethical. A very sad example of this occurred in the National Surgical Adjuvant Breast and Bowel Project (NSABP): This multi-center study demonstrated that lumpectomy could be equivalent to mastectomy in hundreds of thousands of women. The Chairman of this study discovered that the Principal Investigator in one of the clinical centers had falsified some patient records so that women who were not eligible to be in the study based on pre-determined enrollment criteria, were made falsely eligible to participate. This excellent and extremely important study was initially tainted when this became known and the Chairman of the study was charged by the Office of Research Integrity (ORI) with scientific misconduct, even though he had notified the NIH of the problem when he learned of it. He was subsequently completely cleared, and he was offered multiple apologies. The study has had profound implications on the treatment of women with breast cancer. Nevertheless, this was a serious breach of ethics on the part of an investigator in one of the many centers, that could have invalidated the findings. Fortunately the results held up even when all the patients from the offending clinic were excluded.

9.6 Authorship policies

In medical research most original research articles have multiple authors, since medical research is a collaborative effort. Most medical journals, and research institutions, have specific and strict authorship policies (published in journals and/or on websites) many of which embody the following elements: (1) co-authors must make an intellectual contribution to the paper (e.g. conceive the research, perform analyses, write sections of the paper, or make editorial contributions); (2) all co-authors must bear responsibility for its contents; (3) co-authors must disclose potential conflicts of interest (e.g. relevant support from industry, lectureships, stock ownership). Order of authorship may sometimes be a point of contention and should be discussed by the co-authors early in the process.

9.7   Data and Safety Monitoring Boards

Generally, clinical trials have a Data and Safety Monitoring Board (DSMB) to oversee the trial. These are independent groups of experts in the relevant disciplines who are in an advisory capacity. Their job is to monitor the trial and to assure the safety of participants. In a blinded trial they are the only ones who see the unblinded data at regular, pre-specified intervals. If they find excessive benefit or harm in one arm of the trial, they would advise to stop the trial (as happened in the Women's Health Initiative). Usually the criteria for stopping a trial due to harm in the treatment group are more stringent than stopping for benefit.

9.8   Summary

The ethical conduct of research has many components. New and difficult ethical questions arise as science advances and new technologies become available. This brief chapter just begins to give you an idea of some of the issues involved. Much more detailed information is available from various websites and NIH has an on-line course in protection of human subjects. Local IRB's can you give you information and additional sources.

Postscript
A FEW PARTING COMMENTS
ON THE IMPACT OF EPIDEMIOLOGY
ON HUMAN LIVES

Ten years ago a woman with breast cancer would be likely to have a radical mastectomy, which in addition to removal of the breast and the resulting disfigurement, would also include removal of much of the muscle wall in her chest and leave her incapacitated in many ways. Today, hardly anyone gets a radical mastectomy and many don't even get a modified mastectomy, but, depending on the cancer, may get a lumpectomy which just removes the lump, leaving the breast intact. Years ago, no one paid much attention to radon, an inert gas released from the soil and dissipated through foundation cracks into homes. Now it is recognized as a leading cause of lung cancer. The role of nutrition in prevention of disease was not recognized by the scientific community. In fact, people who believed in the importance of nutrients in the cause and cure of disease were thought to be faddists, just a bit nutty. Now it is frequently the subject of articles, books, and news items, and substantial sums of research monies are invested in nutritional studies. Such studies influence legislation, as for example the regulations that processed foods must have standard labeling, easily understood by the public at large, of the fat content of the food as well as of sodium, vitamins, and other nutrients. All this has an impact on the changing eating habits of the population, as well as on the economics of the food industry.

In the health field changes in treatment, prevention, and prevailing knowledge come about when there is a confluence of circumstances: new information is acquired to supplant existing theories; there is dissemination of this information to the scientific community and to the public at large; and there is the appropriate psychological, economic, and political climate that would welcome the adoption of the new approaches. Epidemiology plays a major role by providing the methods by which new scientific knowledge is acquired. Often, the first clues to causality come long before a biological mechanism is known. Around

197

1850 in London, Dr. John Snow, dismayed at the suffering and deaths caused by epidemics of cholera, carefully studied reports of such epidemics and noted that cholera was much more likely to occur in certain parts of London than in other parts. He mapped the places where cholera was rampant and where it was less so, and he noted that houses supplied with water by one company, the Southwark and Vauxhall Company, had many more cases of cholera than those supplied by another company. He also knew that the Vauxhall Company used as its source an area heavily contaminated by sewage. Snow insisted that the city close the pump supplying the contaminated water, known as the Broad Street Pump. They did so and cholera abated. All this was 25 years before anyone isolated the cholera bacillus and long before people accepted the notion that disease could be spread by water. In modern times, the AIDS epidemic is one where the method of spread was identified before the infectious agent, the HIV virus, was known.

Epidemiologic techniques have been increasingly applied to chronic diseases, which differ from infectious diseases in that they may persist for a long time (whereas infections usually either kill quickly or are cured quickly) and also usually have multiple causes, many of which are difficult to identify. Here, also, epidemiology plays a central role in identifying risk factors, such as smoking for lung cancer. Such knowledge is translated into public action before the full biological pathways are elucidated. The action takes the form of educational campaigns, anti-smoking laws, restrictions on advertisement, and other mechanisms to limit smoking. The risk factors for heart disease have been identified through classic epidemiologic studies resulting in lifestyle changes for individuals as well as public policy consequences.

Chronic diseases present different and challenging problems in analysis, and new statistical techniques continue to be developed to accommodate such problems. New statistical techniques are also being developed for the special problems encountered in genetics research. Thus the field of statistic is not static and the field of epidemiology is not fixed. Both adapt and expand to deal with the changing health problems of our society.

Appendix A
CRITICAL VALUES OF CHI-SQUARE, Z, AND t

When Z, χ^2, or t value calculated from the observed data is equal to or exceeds the critical value listed below, we can reject the null hypothesis at the given significance level, α (alpha).

Selected Critical Values of Chi-Square

Significance Level	.1	.05	.01	.001
Critical Value of χ^2	2.71	3.84	6.63	10.83

Selected Critical Values of Z

Significance Level Two-Tailed Test (One-Tailed Test)	.1 (.05)	.05 (.025)	.01 (.005)	.001 (.0005)
Critical Value of Z	1.64	1.96	2.58	3.29

Selected Critical Values of t

Significance Level Two-Tailed Test (One-Tailed Test)	.10 (.05)	.05 (.025)	.01 (.005)	.001 (.0005)
Degrees of Freedom				
9	1.83	2.26	3.25	4.78
19	1.73	2.09	3.86	3.88
100	1.66	1.98	2.63	3.39
1000	1.64	1.96	2.58	3.29

NOTE: Interpretation:
If you have 19 degrees of freedom, to reject H_o, at $\alpha = .05$ with a two-tailed test, you would need a value of t as large or larger than 2.09; for $\alpha = .01$, a t at least as large as 3.86 would be needed. Note that when df gets very large the critical values of t are the same as the critical values of Z. Values other than those calculated here appear in most of the texts shown in the Suggested Readings.

Appendix B
FISHER'S EXACT TEST

Suppose you have a 2-by-2 table arising from an experiment on rats that exposes one group to a particular experimental condition and the other group to a control condition, with the outcome measure of being alive after one week. The table looks as follows:

Table B.1

	Control	Experimental	
Alive	a 1	b 7	Row $1 = R_1 = 8$
Dead	c 5	d 1	Row $2 = R_2 = 6$
Total	Col $1 = C_1 = 6$	Col $2 = C_2 = 8$	$N = 14$

87.5% of the experimental group and 16.7% of the control group lived. A more extreme outcome, given the same row and column totals, would be

Table B.2

	Control	Experimental	
Alive	0	8	8
Dead	6	0	6
Total	6	8	14

where 100% of the experimental and 0% of the control group lived. Another more extreme outcome would be where 25% of the experimental and all of the controls lived:

201

Table B.3

	Control	Experimental	
Alive	6	2	8
Dead	0	6	6
Total	6	8	14

(Any other tables we could construct with the same marginal totals would be less extreme than Table B.1, since no cell would contain a number less than the smallest number in Table B.1, which is 1.)

We calculate the exact probability of getting the observed outcome of the experiment by chance alone (Table B.1), or one even more extreme (as in either Table B.2 or B.3), if it were really true that there were no differences in survival between the two groups. Fisher's exact test is calculated by getting the probability of each of these tables and summing these probabilities.

First we have to explain the symbol "!". It is called a "factorial." A number $n!$ means $(n) \times (n–1) \times (n–2) \times \ldots \times (1)$. For example, $6! = 6 \times 5 \times 4 \times 3 \times 2 \times 1$. By definition, $0!$ is equal to 1.

The probability of getting the observations in Table B.1 is

$$\frac{(R_1!) \times (R_2!) \times (C_1!) \times (C_2!)}{a! \times b! \times c! \times d! \times N!} = \frac{8!6!6!8!}{1!7!5!1!14!} = .015984$$

The probability of getting the observations in Table B.2 is

$$\frac{8!6!6!8!}{0!8!6!0!14!} = .000333$$

The probability of getting the observations in Table B.3 is

$$\frac{8!6!6!8!}{6!2!0!6!14!} = .009324$$

The sum of these probabilities is $.015984 + .000333 + .009324 = .025641$. Thus we can say that the exact probability of obtaining the results we observed in Table B.1, or results more extreme, is .025641, if the null hypothesis was true. We may reject the null hypothesis that the survival rate is the same in both groups at a significance level $p = .026$.

Appendix C
KRUSKAL–WALLIS NONPARAMETRIC TEST TO COMPARE SEVERAL GROUPS

For example, suppose you have three groups of people each having a score on some scale. The total number of people in all three groups is N. The general procedure is as follows: (1) Combine all the scores from the three groups and order them from lowest to highest. (2) Give the rank of 1 to the lowest score, 2 to the next lowest, and so on, with N being assigned to the person with the highest score. (3) Sort the people back into their original groups, with each person having his assigned rank. (4) Sum all the ranks in each group. (5) Calculate the quantity shown below, which we call H. (6) If you have more than five cases in each group, you can look up H in a chi-square table, with $k-1$ degree of freedom (where k is the number of groups being compared).

Scores On Reading Comprehension					
Group A		Group B		Group C	
Scores	(Rank)	Scores	(Rank)	Scores	(Rank)
98	(13)	80	(9)	120	(21)
70	(6)	60	(2)	110	(17)
68	(5)	106	(15)	90	(12)
107	(16)	50	(1)	114	(19)
115	(20)	75	(8)	105	(14)
65	(4)	74	(7)	85	(10)
Sum	**(64)**	64	(3)	112	(18)
of			**(45)**	87	(11)
Ranks					**(122)**

$$H = \left(\frac{12}{N(N + 1)} \times \Sigma \frac{(R)^2}{n_j} \right) - 3(N + 1)$$

$$H = \left(\frac{12}{21(22)} \right) * \left(\frac{64^2}{6} + \frac{45^2}{7} + \frac{122^2}{8} \right) - 3(22)$$

$$= 7.57$$

degrees of freedom = 3–1 = 2

If the null hypothesis of no difference in mean rank between groups was true, the probability of getting a chi-square as large as, or larger than, 7.57 with 2 degrees of freedom is less than .05, so we can reject the null hypothesis and conclude the groups differ. (When ties occur in ranking, each score is given the mean of the rank for which it is tied. If there are many ties, a correction to H may be used, as described in the book by Siegel listed in Suggested Readings.)

Appendix D
HOW TO CALCULATE A
CORRELATION COEFFICIENT

Individual	X	Y	X^2	Y^2	XY
A	5	7	25	49	35
B	8	4	64	16	32
C	15	8	225	64	120
D	20	10	400	100	200
E	25	14	625	196	350
Σ	73	43	1339	425	737

$$r = \frac{N\Sigma XY - (\Sigma X)(\Sigma Y)}{\sqrt{N\Sigma X^2 - (\Sigma X)^2}\ \sqrt{N\Sigma Y^2 - (\Sigma Y)^2}}$$

$$= \frac{5(737) - (73)(43)}{\sqrt{5(1339) - (73)^2}\ \sqrt{5(425) - (43)^2}} = \frac{3685 - 3139}{\sqrt{1366}\ \sqrt{276}}$$

$$= \frac{546}{(37)(16.6)} = \frac{546}{614} = .89$$

How to Calculate Regression Coefficients:

$$b = \frac{\Sigma XY - \dfrac{(\Sigma X)(\Sigma Y)}{N}}{\Sigma X^2 - \dfrac{(\Sigma X)^2}{N}}\ ; \quad a = \Sigma\frac{Y}{N} - b\frac{\Sigma X}{N}4$$

$$b = \frac{737 - \dfrac{(73)\,(43)}{5}}{1339 - \dfrac{(73)^2}{5}} = \frac{737 - 628}{1339 - 1066} = \frac{109}{273} = .40$$

$$a = \frac{43}{5} - \frac{.40\,(73)}{5} = 8.60 - 5.84 = 2.76$$

Appendix E
AGE-ADJUSTMENT

Consider two populations, A and B, with the following characteristics:

Population	Age	Age-Specific Rates	# of People in Population	# of Deaths in Population	Crude Death Rate
A	Young	$\dfrac{4}{1000} = .004$	500	$.004 \times 500 = 2$	
	Old	$\dfrac{16}{1000} = .016$	$\underline{500}$	$.016 \times 500 = \underline{8}$	
	Total		1000	10	$\dfrac{10}{1000}$
B	Young	$\dfrac{5}{1000} = .005$	667	$.005 \times 667 = 3.335$	
	Old	$\dfrac{20}{1000} = .020$	$\underline{333}$	$.020 \times 333 = \underline{6.665}$	
	Total		1000	10	$\dfrac{10}{1000}$

Note that the population B has higher age-specific death rates in each age group than population A, but both populations have the same crude death rate of 10/1000. The reason for this is that population A has a greater proportion of old people (50%) and even though the death rate for the old people is 16/1000 in population A compared with 20/1000 in population B, the greater number of people in that group contribute to a greater number of total deaths.

To perform age adjustment, we must select a standard population to which we will compare both A and B. The following examples use two different standard populations as illustrations. In practice, a standard population is chosen either as the population during a particular year or as the combined A and B population. The choice of standard population does not matter. The phrase "standard population" in this context refers to a population with a particular age distribution (if we are adjusting for age) or sex distribution (if we are adjusting for sex). The age-specific (or sex-specific, if that is what is being adjusted) rates for both group A and B are applied to the age distribution of the standard population in order to compare A and B *as if* they had the same age distribution.

Note if you use two different standard populations you get different age-adjusted rates but relative figures are the same, that is, the age-adjusted rates for A are lower than for B. This implies that the age-specific rates for A are lower than for B, but since the crude rates are the same it must mean that population A is older. Because we know that age-specific rates for older people are higher than for younger people, population A must have been weighted by a larger proportion of older people (who contributed more deaths) in order to result in the same crude rate as B but in a lower age-adjusted rate.

There are exceptions to the above inference when we consider groups where infant mortality is very high. In that case it could be that the young have very high death rates, even higher than the old. In industrialized societies, however, the age-specific death rates for the old are higher than for the young.

STANDARD POPULATION I: Example (more old people than young)

Age	# of People	Apply Age-Specific Death Rates for Population A to Standard Population	# of Deaths Expected in A if It Were the Same Age Composition as the Standard Population	Apply Age-Specific Death Rates for Population B to Standard Population	# of Deaths Expected in B if It Were the Same Age Composition as the Standard Population
Young	300	× .004 =	1.2	.005	1.5
Old	700	× .016 =	11.2	.020	14.0
Total	1000		12.4		15.5
Age-adjusted rates for: A = 12/1000 B = 15/1000					

STANDARD POPULATION II: Example (more young people than old)

Age	# of People	Apply Age-Specific Death Rates for Population A to Standard Population	# of Deaths Expected in A if It Were the Same Age Composition as the Standard Population	Apply Age-Specific Death Rates for Population B to Standard Population	# of Deaths Expected in B if It Were the Same Age Composition as the Standard Population
Young	1167	× .004 =	4.67	.005	5.84
Old	833	× .016 =	13.33	.020	16.66
Total	2000		18		22.50

Age-adjusted rates for:	Age-adjusted rates for:
$A = \dfrac{18}{2000}$	$B = \dfrac{22.5}{2000}$
$= \dfrac{9}{1000}$	$= \dfrac{11.25}{1000}$

Appendix F
CONFIDENCE LIMITS ON ODDS RATIOS

The 95% confidence limits for an odds ratio (OR) are

$$OR \times e^{\left[\pm 1.96\sqrt{\frac{1}{a} + \frac{1}{b} + \frac{1}{c} + \frac{1}{d}}\right]}$$

We reproduce here the table from Section 4.12 to use as an example:

	Patients with Lung Cancer		Matched Controls with Other Diseases	
Smokers of 14–24 Cigarettes Daily	475	a	431	b
Nonsmokers	7	c	61	d

	(persons with disease)	(persons without disease)

$$OR = \frac{475 \times 61}{431 \times 7} = 9.6$$

Upper 95% confidence limit =

$$OR \times e^{\left[1.96\sqrt{\frac{1}{475} + \frac{1}{431} + \frac{1}{7} + \frac{1}{61}}\right]}$$

$$OR \times e^{(1.96 \times .405)} = 9.6 \times e^{.794} = 9.6 \times 2.21 = 21.2$$

Lower 95% confidence limit =

$$OR \times e^{(-1.96 \times .405)} = 9.6 \times e^{-.794} = 9.6 \times .45 = 4.3$$

$$Note: \quad e^{-.794} = \frac{1}{e^{.794}} = .45$$

Thus, the confidence interval is 4.3 to 21.2.

Appendix G
"J" OR "U" SHAPED RELATIONSHIP BETWEEN TWO VARIABLES

Sometimes we may think that a variable is related to an outcome in a J or U shape, for example weight and mortality. (The index of weight we use is the body mass index – BMI – which is weight in kilograms divided by the square of the height in meters). A "J" or "U" curve can be imagined by plotting body mass index against mortality. The J or U shape means that as BMI goes up, the death rate goes up, but the death rate also goes up at very low values of BMI, hence the J or U shape. This may be due to pre-existing illness: people who are on the very thin side may have lost weight because they are already ill and so of course they will be more likely to die. Or it may be due to the physiological consequences of very low weight. We can test whether there is a J curve by including a square term (also known as a quadratic term) in the Cox equation.

The figure below shows an example of a J shaped relationship between BMI and some outcome.

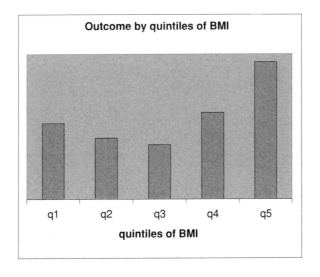

As you can see, there is a U shape to the curve. Both those at lower and higher BMI levels have higher levels of the outcome.

Below are coefficients[1] from SHEP (Systolic Hypertension the Elderly Program). from a Cox regression equation for death among people in the active treatment group. A quadratic term for BMI was entered into the Cox model, along with a number of other covariates, resulting in the following coefficients for BMI:

	Coefficient	Standard error	p value
BMI	−0.3257	0.1229	.008
BMI^2	0.0059	0.0020	.003

Because the *coefficient for BMI squared is significant*, we know the relationship of BMI to mortality is not linear.

1. **To calculate the relative risk of death for a given BMI value compared to another BMI** value we must do the following:

$$RR = e^k \text{ (see Section 4.20)}$$

To compare a BMI of 32 to a BMI of 27 if the coefficient for the square term is significant,
k = linear term coefficient x $(BMI_1 - BMI_2)$ + square term coefficient $(BMI_1^2 - BMI_2^2)$
$k = -0.3257 (32 - 27) + 0.005 (32^2 - 27^2) = .1120$
$e^k = 1.12$
people with a BMI of 32 are 12% more likely to die than those with a BMI of 27.

---

1. Wassertheil-Smoller S, Fann C, Allman RM, Black HR, Camel GH, Davis B, Masaki K, Pressel S, Prineas RJ, Stamler J, Vogt TM, For the SHEP Cooperative Research Group. Relation of low body mass to death and stroke in the Systolic Hypertension in the Elderly Program (SHEP). *The Archives of Internal Medicine* 2000;160(4):494-500.

{Note: if the square term coefficient were not significant, then k = linear coefficient [$BMI_1 - BMI_2$]}

2. **To calculate the nadir (lowest point) of the J or U curve:**

NADIR $= -1/2$(linear coefficient/quadratic coefficient)
$= -1/2 (-0.3257/0.0059)$
$= -1/2 \times (-55.20) = -27.6$ (rounded)

This means that the lowest mortality for this group occurs at a BMI of 27.6. Because of rounding errors the actual nadir reported in the reference pages was 27.7.

3. **Now we may want to get the risk of a given BMI relative to the nadir.**
   a) RR of BMI = 25 compared to nadir calculated as
      $-0.3257 (25 - 27.7) + .0059 (25^2 - 27.7^2) = .0399$
      $e^k = 1.04$
   b) RR of BMI = 30.4
      $K = -.3257 (30.4 - 27.7) + .0059 (30.4^2 - 27.7^2) = .0461$
      $e^k = 1.05$
   c) RR of BMI below nadir and above nadir is elevated

Appendix H
DETERMINING APPROPRIATENESS OF CHANGE SCORES

(1) To determine if change scores are appropriate:

Consider a group of 16 patients who have the following scores on a scale assessing depressive symptoms; a retest is given shortly after to determine the variability of scores within patients:

Table H.1

Patient #	First Test Scale Score	Retest Score
1	12	13
2	16	15
3	22	21
4	24	23
5	30	29
6	18	19
7	16	15
8	12	12
9	14	15
10	18	18
11	24	24
12	30	29
13	18	19
14	16	15
15	14	15
16	10	11
Mean	18.38	18.31

An analysis of variance indicates the following (Table H.2):

Table H.2

Source of Variation	SS	df	MS	F	*P*-value
Patients	1014.7188	15	67.6479	156.8647	0.0000
Test retest	0.0313	1	0.0313	0.0725	0.7915
Error	6.4688	15	0.4313		
Total	1021.2188	31			

$\sigma^2_{\text{between patients}} = (67.65 - .43)/2 = 33.61$

$\sigma^2_{\text{between patients + error}}\ 33.61/(33.61 + .43) = .987$

This is greater than .5, so that the use of change scores is appropriate. [Note: σ^2, or the variance, is the MS (mean square) from the analysis of variance.]

Next, the patients are divided into two groups; one group is given a dietary intervention lasting 10 weeks, while the other group serves as a control group. The scale is administered again after 10 weeks to both groups, with the following results:

Table H.3

	Control Group				Treatment Group		
Patient #	Pre-test	Post-test	Change Score	Patient #	Pre-test	Post-test	Change Score
1	12	13	1	9	14	7	−7
2	16	15	−1	10	18	10	−8
3	22	20	−2	11	24	7	−17
4	24	18	−6	12	30	5	−25
5	30	25	−5	13	18	10	−8
6	18	16	−2	14	16	8	−8
7	16	12	−4	15	14	4	−10
8	12	10	−2	16	10	5	−5
Mean	18.75	16.13	−2.63		18	7	−11
Variance	38.79	23.27	5.13		40.00	5.14	44.57
s.d.	6.23	4.82	2.26		6.32	2.27	6.68

(2) To calculate *coefficient of sensitivity* to change, do a repeated measures analysis of variance on the scores in the *treatment group*; to get the error variance, calculate the variance of the change scores.

Table H.4

Source of Variation	SS	df	MS	F	P value
Between test/retest	484	1	484.0000	21.4430	0.0004
Within patients	316	14	22.5714		
Total	800	15			

Coefficient of sensitivity = variance of change scores in treatment group/(variance of change scores + error variance) =

$$\frac{44.57}{(44.57+22.29)} = .67$$

(the 44.57 is obtained from in last column of Table G.3).

(3) *Effect size* = mean of the change scores/ s.d. of pretest scores:

in the treatment group = $-11/6.32 = -1.74$ (there was a *decline in depression symptom score* of 1.74 pretest standard deviation units).

(4) *Guyatt's responsiveness measure*[25] is (mean change scores in the treatment group)/(s.d. of change scores in stable subjects). We are assuming here that the control group is the group of stable subjects, although generally "stable subjects" refers to subjects who are stable with regard to some external criterion.

$$G = \frac{-11}{2.26} = -4.86$$

(5) *Comparison with a control group:* The effect size for the treatment group is -1.74, so clearly it exceeds the control group change, which is $-2.63/6.23 = -.42$. If we calculate the ratios of treatment to control group for the above indices of responsiveness, we will find in this example that they are very similar.

For effect size the ratio is $-1.74/-.42 = 4.14$. For Guyatt's statistic it is $-4.86/-1.16 = 4.19$. (The -1.16 was obtained by mean change in control group divided by standard deviation of change scores in control group; i.e., $= -2.63/2.26$.)

For the coefficient of sensitivity, it is $.67/.14 = 4.78$. (The .14 was obtained by doing an analysis of scores in the control group, not shown here, so take it on faith, or calculate it as a check on the accuracy of this.)

Appendix I
GENETIC PRINCIPLES

Everything should be made as simple as possible,
but not one bit simpler.

(attributed to Albert Einstein, 1879–1955)

1. DNA (deoxyribonucleic acid) is made up of four units—or nucleo-tides—that are molecules composed of carbon, hydrogen, oxygen and phosphorous. These molecules, called bases, are adenine, guanine, thymine and cytosine, and are denoted by the letters A, G, T and C.

2. The DNA is arranged in two strands twisted in a double helix form, such that the nucleotides AGCT pair with each other in fixed ways. An A always pairs with T and C always pairs with G. These are called base pairs. If one strand of the double helix were strung out in a line it might look like this:

AATTCGTCAGTCCC, The other strand that pairs with it would be: TTAAGCAGTCAGGG.

3. There are 3 billion such base pairs (or 6 billion bases) in the hu-man genome (which refers to all the genetic material in humans). These 3 billion base pairs are organized into 23 chromosome pairs (one from the mother and one from the father), which are in every living cell in the body.

4. Out of these 3 billion base pairs there are about 30,000 genes (the estimate varies—in 1970 it was 300,000; in the 1990's it was 100,000; subsequently, after the decoding of the human genome, it was thought to be 30,000; more recently, the estimate has been upped again to maybe 45,000), which are sequences of base pairs of different lengths. The remaining sequences are mostly "junk" DNA, though some sequences, not genes in themselves, serve regulatory functions.

221

5.   Genes instruct the cell how to make proteins (or polypeptide chains) that are composed of amino acids.

6.   A group of 3 bases is called a codon. A codon codes for a single amino acid. Proteins consist of several or many amino acids. Since there are 4 different letters (nucleotides), one would think there should be 4x4x4x4 or 64 different amino acids, but there are only 20 (don't ask why). The basis of all life is these 20 amino acids.

7.   Genes make proteins by "coding" for amino acids. This means that the gene directs the assembly of the protein molecule by specifying the sequence of amino acids and how they should bind to make the particular protein, an enzyme for example that converts cholesterol to estrogen or a neurotransmitter, or bone.

8.   A bunch of codons (each made up of 3 bases and each coding for a specific one of the 20 amino acids) is called an exon. Thus, an exon is composed of many codons, each coding for a single amino acid, and together coding for a specific protein.

9.   In between exons in a gene there are other sequences of letters called introns. Introns are sections of DNA (stretches of the letters A,G,C,T) which do not code for amino acids. It is not known what the functions of introns are, if any.

10. A gene then, consists of exons and introns. Before it can actually direct the manufacture of a protein, it is transcribed within the nucleus of the cell into RNA (ribonucleic acid) which is an exact copy of a single strand of the DNA, (except that instead of the base thymine, T, it contains uracil, U). This RNA then gets out of the nucleus and into the body (cytoplasm) of the cell where the proteins will be manufactured. The incredible thing is that then the parts of the RNA that are not exons (or coding regions), are removed and this is now called mRNA (messenger RNA). So at the end of this transcription process, all the introns are spliced out and only the exons (coding regions) remain, ready to direct the materials within the cell, outside of the nucleus, to manufacture the chain of amino acids which constitute the particular protein coded by that particular gene. Once the protein is manufactured, it is either used by the cell or transported in vesicles to the membrane of the cell where it is expelled into the tissue or blood.

11. When we say a gene is "expressed" it means that mRNA and a protein is made. Only about 3% of the genome (i.e. 3% of all the DNA) is expressed; the rest is either regulatory or is junk (as far as we know now).

12. Since not all of the DNA is composed of genes, finding genes embedded in all that DNA is a major challenge. But let us say you find a sequence of bases that is ATG, which codes for the amino acid methionine. This is also known as the Start Codon because it tells the cell to start making its protein from that point on; the cell continues making the amino acids in the sequence specified by the codons, until it encounters the Stop Codon which has one of three forms: TAA, TAG, or TGA, at which point it stops.

13. There are many redundancies in the system, and the third position is "wobbly," meaning for example, that even if instead of TCA, you have a TCC, you could still make the same amino acid (serine). On the other hand, it might be that a single change of a letter, say having a T where most people have a C, could mess up the whole protein because an amino acid is missing or a different one is made.

14. These variations between people in one single letter of the code are known as single nucleotide polymorphisms (SNPs, pronounced as snips). It is estimated that there may roughly be one SNP per 1000 bases of DNA. Assume the SNP can come in two forms—i.e. two alleles, meaning a particular spot could have, say, the nucleotide A or a T. If you are looking at a stretch of DNA 2000 bases long, there could be 4 variations of SNPs, 2 in the first 1000 bases and 2 in the second 1000 bases. Some SNPs are normal variants in a population, some may predispose to disease, and some may cause disease.

15. Polymorphisms, then, are variants of a gene or variants at a certain place or locus within a gene, most of which are harmless variants. A particular polymorphism for example may consist of two alternate forms (alleles) at a particular locus; for example you could have either an A or a C at that locus. Polymorphisms may also consist of insertions of a nucleotide at a particular place, or a deletion of a nucleotide, or other variations.

16. Scientists are beginning to look at blocks of DNA—long stretches of DNA at a particular location on a chromosome which have a distinctive pattern of SNPs. These are called haplotypes. Haplotypes of interest are sequences of alleles that are close together on the same chromosome and thus tend to be inherited as a block. It may be possible to look at the association with disease of the SNPs that differentiate some common patterns of haplotypes, rather than looking at the association of each individual SNP with disease. Construction of haplotype maps ("hapmaps") is an active area of research that shows much promise in being able to identify disease polymorphisms more efficiently.

17. The whole human genome then consists of 30,000 or so genes plus all the other genetic material which have no, or unknown, function, which make up the 3 billion base pairs. Each and every nucleated cell in the body has the same 3 billion base pairs (except the sperm and the egg, each of which have half of that). So the question might be if each cell has the same DNA why doesn't each cell make the same proteins? Why doesn't a liver cell make insulin? Why doesn't a brain cell make bone? That's because part of a gene, or sometimes a sequence at some distance away from the gene, tells the gene when to turn on or off—that is, it regulates the expression of the gene.

18. During cell division (mitosis), in all cells except the sperm and egg, there is first, a duplication of the chromosomes and then a segregation of the duplicated parts so that when the cell splits in half the two new daughter cells will each have a complete set of 23 pairs of chromosomes. The two members of each pair carry copies of the same genes (and are homologous), but may differ in certain spots (loci) on the chromosome because one member of the pair comes from the mother and one from the father.

19. In the case of germ cells in the testes or ovaries, which are destined to become sperm and egg, the cell division is called meiosis, and in this kind of cell division, there is no duplication of the chromosomes before division, so that when sperm and egg meet, only one set of chromosomes results (one chromosome of each pair in the offspring comes from the sperm and one from the egg). During

meiosis, the two homologous chromosomes in the chromosome pair come very close to each other and little bits of DNA cross over from one chromosome of the pair to the other chromosome of the pair. Now remember this is in the germ cell (the precursor cell of the sperm and egg). This germ cell than splits (by meiosis, rather than mitosis) So when the sperm is finally formed, it has only one member of each pair of chromosomes, but this one chromosome of the pair is just a little different than the original chromosome which is in every cell of the parent body, because of the exchange of some of the DNA that took place during meiosis. (This is called recombination). Something similar happens to the egg, and when the two meet in fertilization and form a gamete (a single fertilized cell from which the human organism will develop) this cell now has a full set of 23 pairs of chromosomes, one in each pair from the mother and one from the father.

20. Recombination is important in the discovery of genes related to disease because of the following: if the alleles at two loci on the chromosome are close together, they will be inherited together but if they are far enough apart, they are more likely to recombine (cross over) during meiosis and will be inherited independently of each other. This state of affairs helps scientists to locate genes. Recombination due to the crossing over of part of the DNA from one chromosome of a pair to the other, happens only during meiosis in the germ cell. Thereafter, every time a cell reproduces, and they do all the time, it splits in two with exactly the same copies of the complement of 23 chromosome pairs.

21. Mutations are errors in the replication of the DNA. They arise spontaneously very rarely—about 1 in 1,000,000,000 bases, but environmental exposures may cause mutations. Mutations may be somatic, i.e. in cells of the living organisms, in which case they die out when the organism dies, or in germ cells which go on to be sperm or egg and in which case they get transmitted on through the generations. Mutations may be of several types and they may have no effect, for example if they happen in an intron, or they may change the amino acid being produced and thus may alter the protein. A mutation may delete a single nucleotide, insert a single nu-

cleotide or substitute one nucleotide for another, say substituting a C for a T. An insertion or deletion mutation may cause a "frame shift," meaning that the start and/or end of the coding region would be different and so a different sequence of amino acids would be formed. In a "missense mutation" one of the 20 amino acids is replaced by another (thereby changing the protein); in a "nonsense" mutation, a stop codon appears prematurely. As noted before, some mutations have no effect on the organism and some may cause or pre-dispose to disease.

In fact, it's miraculous that when so much can go wrong, so little actually does!

REFERENCES

Chapter 1

1. Popper KR: The Logic of Scientific Discovery. New York: Harper and Row, 1959.
2. Weed DL: On the Logic of Causal Inference. American Journal of Epidemiology, 123(6):965–979, 1985.
3. Goodman SN, Royall R: Evidence and Scientific Research. AJPH, 78(12):1568–1574, 1988.
4. Susser M: Rules of Inference in Epidemiology. In: Regulatory Toxicology and Pharmacology, Chapter 6, pp. 116–128. New York: Academic Press, 1986.
5. Brown HI: Perception, Theory and Commitment: The New Philosophy of Science. Chicago: Precedent, 1977.
6. Hill B: Principles of Medical Statistics, 9th ed. New York: Oxford University Press, 1971.

Chapter 2

7. Arias E, Smith BL. Deaths: Preliminary Data for 2001. National vital statistics reports; vol. 51, no. 5. Hyattsville, Maryland: National Center for Health Statistics. 2003.

Chapter 3

8. Drapkin A, Mersky C: Anticoagulant Therapy after Acute Myocardial Infarction. JAMA, 222:541–548, 1972.
9. Intellectual Development of Children: U.S. Department of Health, Education, and Welfare, Public Health Service, HSMHA, December 1971, Vital and Health Statistics Series 11, November 10.

10. Davis BR, Blaufox MD, Hawkins CM, Langford HG, Oberman A, Swencionis C, Wassertheil-Smoller S, Wylie-Rosett J, Zimbaldi N: Trial of Antihypertensive Interventions and Management. Design, Methods, and Selected Baseline Results. Controlled Clinical Trials, 10:11–30, 1989.
11. Oberman A, Wassertheil-Smoller S, Langford H, Blaufox MD, Davis BR, Blaszkowski T, Zimbaldi N, Hawkins CM, for the TAIM Group: Pharmacologic and Nutritional Treatment of Mild Hypertension: Changes in Cardiovascular Risk Status. Annals of Internal Medicine, January, 1990.
12. Rothman KJ: No Adjustments are Needed for Multiple Comparisons. Epidemiology, 1(1):43–46, 1990.
13. Scarr-Salapateck S: Race, Social Class, and I.Q. Science, 174: 1285–1295, 1971.
14. Sokal RR, Rohlf JF: Biometry. San Francisco: W.H. Freeman, 1969.

Chapter 4

15. NCHS: Advanced Report of Final Mortality Statistics, 1987, Supl. (p. 12), MSVR Vol. 38, No. 5, Pub. No. (PHS) 89-1120, Hyattsville, MD: Public Health Service, September 26, 1989.
16. Hypertension Detection and Follow-Up Program Cooperative Group: Blood Pressure Studies in 14 Communities. A Two-Stage Screen for Hypertension. JAMA, 237(22):2385–2391, 1977.
17. Inter-Society Commission for Heart Disease Resources. Atherosclerosis Study Group and Epidemiology Study Group: Primary Prevention of the Atherosclerotic Diseases. Circulation, 42:A55, 1970.
18. Kannel WB, Gordon T, eds. The Framingham Study: An Epidemiological Investigation of Cardiovascular Disease. Feb. 1974, DHEW Publ. No. (NIH) 74–599.
19. These data come from the National Pooling Project. For purposes of this example, high blood pressure is defined as diastolic blood pressure $\geq$ 105 mm Hg and "normal" is defined as PB <

78 mm Hg. The disease in question is a "Coronary Event" and the time periods is 10 years. Note that hypertension is currently defined as systolic blood pressure $\geq$ 140 mmHg and/or diastolic blood pressure $\geq$ 90 mmHg. Note: current bounds for hypertension are a systolic blood pressure $\geq$ 140 mmHg or a dialostic blood pressure $\geq$ 90 mmHg.

20. Lilienfeld AM: Foundations of Epidemiological, p. 180. New York: Oxford University Press, 1976.
21. Writing Group for The Women's Health Initiative Investigators. Risks and Benefits of Estrogen Plus Progestin in Healthy Post-menopausal Women. JAMA-Express, 288:321-333, 2002.
22. D'Agostino Jr. RB. Tutorial in Biostatistics Propensity Score Methods for Bias Reduction in the Comparison of a Treatment to a Non-Randomized Control Group. Statistics in Medicine, 17:2265-2281, 1998.
23. The numerical example is courtesy of Dr. Martin Lesser, Cornell University Medical Center.
24. Greenland S. Modeling and Variable Selection in Epidemiologic Analysis. AJPH, 79(3):340-349.

Chapter 5

25. Meyer KB, Pauker SG: Screening for HIV: Can We Afford the False Positive Rate? New England Journal of Medicine, 317(4): - 238–241, 1987.

Chapter 6

26. Coronary Drug Project Research Group: Influence of Adherence to Treatment and Response of Cholesterol on Mortality in the Coronary Drug Project. New England Journal of Medicine, 303:1038–1041, 1980.
27. Peto R, Pike MC, Armitage P, Breslow NE, Cox DR, Howard SV, Mantel N, McPherson K, Peto J, Smith PG: Design and Analysis of Randomized Clinical Trials Requiring Prolonged

Observation of Each Patient. I. Introduction and Design. British Journal of Cancer, 34:585–612, 1976.

Chapter 7

28. Guyatt G, Walter S, Norman G: Measuring Change Over Time: Assessing the Usefulness of Evaluative Instruments. Journal of Chronic Diseases, 40:171–178, 1987.

Chapter 8

29. Ellsworth DL, Manolio TA. The Emerging Importance of Genetics in Epidemiologic Research. I. Basic Conepts in Human Genetics and Laboratory Technology. Ann Epidemiol., 9:1-16, 1999.
30. Ellsworth DL, Manolio TA. The Emerging Importance of Genetics in Epidemiologic Research II. Issues in Study Design and Gene Mapping. Ann Epidemiol. 9:75-90, 1999.
31. Ellsworth DL, Manolio TA. The Emerging Importance of Genetics in Epidemiologic Research. III. Bioinformatics and Statistical Genetic Methods. Ann Epidemiol., 9:207-224, 1999.
32. Lichtenstein P, Holm NV, Verkasalo PK, Iliadou A, Kaprio J, Koskenvuo M, et al. Environmental and heritable factors in the causation of cancer—analyses of cohorts of twins from Sweden, Denmark, and Finland. N Engl J Med 2000;343(2):78-85.
33. Plomin R, Owen MJ, McGuffin P. The Genetic Basis of Complex Human Behaviors. Science, 264:1733-1739, 1994.
34. Risch N, Merikangas K. The future of genetic studies of complex human diseases. Science 1996;273(5281):1516-7.
35. Owen MJ, Cardno AG and O'Donovan MC. Psychiatric genetics: back to the future. Molecular Psychiatry 2000;5:22-31.
36. Risch NJ. Searching for genetic determinants in the new millennium. Nature, 405:847-856, 2000.
37. Burmeister M. Complex Genetics and Implications for Psychiatry. Basic Concepts in the Study of Diseases with Complex Genetics. Biol Psychiatry, 45:522-532, 1999.

38. Schork NJ. Genetics of Complex Disease. Approaches, Problems, and Solutions. Am J Respir Crit Care Med. 156:S103-S109, 1997.

SUGGESTED READINGS

Anderson S, Auquier A, Hauck WW, Oakes D, Vandaele W, Weisberg HI: Statistical Methods for Comparative Studies. New York: John Wiley and Sons, 1980.
> Intermediate level. Requires a prior course in statistics. Excellent for applied researchers.

Balding DJ, Bishop M, Cannings C, editors. Handbook of Statistical Genetics. West Sussex, England, John Wiley & Sons, publishers, 2001.
> A complex reference book.

Campbell DT, Cook TD: Quasiexperimentation: Design and Analysis for Field Settings. Chicago: Rand McNally, 1979.
> Focuses on the design and measurement issues, particularly types of bias, that can arise in quasi-experimental studies. It also deals with the appropriate statistical techniques to be used. Particularly relevant to persons interested in program evaluation.

Cohen J: Statistical Power Analysis for the Behavioral Sciences, 2nd ed. Hillsdale, NJ: Lawrence Erlbaum, 1988.
> A classic and a must. Everything you ever wanted to know about sample size calculations, clearly and comprehensively explained. A reference and source book. Probably needs to be interpreted by a statistician.

Cohen P (ed), Cohen J, West SG, Aiken LS: Applied Multiple Regression: Correlation Analysis for the Behavioral Science, 3rd ed., Lawrence Erlbaum, 2002
> Excellent, thorough, and clear exposition of very complex topics. It's fairly advanced.

Elwood JM: Causal Relationships in Medicine. A Practical System for Critical Appraisal. New York: Oxford University Press, 1988.
    Excellent book with clear explanations of study designs, epidemiologic concepts and relevant statistical methods.

Fleiss JL: Statistical Methods for Rates and Proportions, 2nd ed. New York: John Wiley and Sons, 1982.
    An excellent classic, second-level statistics text concerned with the analysis of qualitative or categorical data.

Fleiss JL: The Design and Analysis of Clinical Experiments. New York: John Wiley and Sons, 1986.
    Focuses on the technical aspects of the design and statistical analysis of experimental studies.

Fryer HPC: Concepts and Methods of Experimental Statistics. Boston: Allyn & Bacon, 1966.
    Basic detailed book on statistical methods. Intermediate to higher level. Many formulas and symbols. More in-depth statistics.

Haines JL and Pericak-Vance MA: Approaches to Gene Mapping in Complex Human Disease. New York: Wiley-Liss, 1998.

Hill, B: Statistical Methods in Clinical and Preventive Medicine. Edinburgh and London: E.S. Livingstone, 1962.
    Classic text on the subject.

Hosmer DW, Lemeshow S: Applied Logistic Regression. New York: John Wiley and Sons, 1989.
    Advanced book on model building in logistic regression, requires statistical background.

Ingelfinger JA, Mosteller F, Thibodeau LA, Ware JH: Biostatistics in Clinical Medicine. New York: Macmillan, 1983.
    Mostly aimed at physicians. Intermediate level. Very good explanations of biostatistics. Many examples from research literature.

Kahn HA: An Introduction to Epidemiologic Methods. New York: Oxford University Press, 1983.
> Intermediate level book on epidemiologic and selected statistical methods. Good explanations of life table analysis and multiple regression and multiple logistic functions. Clear explanations of longitudinal studies using person years.

Kelsey JL, Thompson WD, Evans AS: Methods in Observational Epidemiology. New York: Oxford University Press, 1986.
> Extremely useful explanations of issues involved in case-control, retrospective, and prospective studies. A good discussion of matching, stratification, and design issues.

Kleinbaum DG, Kupper LL: Applied Regression Analysis and Other Multi-variable Methods. Belmont, CA: Wadsworth, 1978.
> This text, as the title indicates, deals with multiple regression and allied topics.

Kleinbaum DG, Kupper LL, Morgenstern E: Epidemiologic Research Principles and Quantitative Methods. Belmont, CA: Wadsworth, 1982.
> An advanced text designed primarily for persons conducting observational epidemiologic research. Both design and statistical issues are covered.

Meinert CL in collaboration with Tonascia S. Clinical Trials: Design, Conduct, and Analysis. New York, Oxford: Oxford University Press, 1986.

Ott J. Analysis of Human Genetic Linkage. Johns Hopkins University Press, 1999.

Popcock SJ: Clinical Trials: A Practical Approach. New York: John Wiley and Sons, 1983.
> An excellent book for anyone undertaking a clinical trial.

Riegelman RK: Studying a Study and Testing a Test: How to Read the Medical Literature. Boston: Little, Brown, 1981.
An introduction to epidemiologic methods and principles aimed at clinicians.

Rothman KJ and Greenland S: Modern Epidemiology. 2nd Ed. Philadelphia: Lippincott Williams & Wilkins, 1998.
An advanced text that covers both design and statistical issues. The focus is an observational epidemiologic study and is directed at the researcher more than the clinician.

Sackett DL, Haynes RB, Tugwell P: Clinical Epidemiology: A Basic Science for Clinical Medicine. Boston: Little, Brown, 1985.
The focus is on applications of epidemiology in clinical practice.

Sham P: Statistics in Human Genetics. London, 1998
Siegel S and Castellan Jr. NJ: Nonparametric Statistics for the Behavioral Sciences. 2nd Edition. New York, Toronto, London: McGraw-Hill, Inc., 1988.
A "how-to-do-it" book; an excellent reference; outlines each procedure step-by-step, a classic.

Sokal RR, Rohlf JF: Biometry, 3rd ed. San Francisco: W.H. Freeman, 1995.
Rather complex and comprehensive.

Streiner L, Norman GR: Health Measurement Scales. A Practical Guide to Their Development and Use. New York: Oxford University Press, 1989.
Excellent and understandable book about scale construction and change scores.

Szklo M, Nieto FJ: Epidemiology. Beyond the Basics. Maryland: Aspen Publishers, 2000.
Somewhere between first and second level text, distinguished by its clear style and very readable, and many excellent examples.

INDEX